Semantics of the Probabilistic Typed Lambda Calculus

Dirk Draheim

Semantics of the Probabilistic Typed Lambda Calculus

Markov Chain Semantics, Termination Behavior, and Denotational Semantics

Springer

Dirk Draheim
Large-Scale Systems Group
Tallinn University of Technology
Tallinn, Estonia

ISBN 978-3-642-55197-0 ISBN 978-3-642-55198-7 (eBook)
DOI 10.1007/978-3-642-55198-7

Library of Congress Control Number: 2017932370

Printed on acid-free paper

This Springer imprint is published by Springer Nature
The registered company is Springer-Verlag GmbH Germany
The registered company address is: Heidelberger Platz 3, 14197 Berlin, Germany

Preface

Today's information systems operate in probabilistic environments. Programs need to react to probabilistic events. Therefore, a rigorous understanding of probabilistic program behavior becomes ever more important. Probabilistic programming is relevant in its own right, as a means to implement randomized algorithms. This book takes a foundational approach to the semantics of probabilistic programming. It deals with the probabilistic typed lambda calculus, which is the typed lambda calculus with recursion plus probabilistic choice.

We elaborate a Markov chain semantics for the probabilistic lambda calculus. As part of this operational semantics, we define a reduction semantics and an evaluation semantics in terms of Markov chain hitting probabilities. The Markov chain semantics unlocks probability theory and Markov chain theory to be used in reasoning about probabilistic programs. Also, we introduce the notions of reduction graphs and reduction trees. Reduction graphs and reduction trees are not part of but rather accompany the Markov chain semantics. They unlock results from graph theory. These prove useful, e.g., in reasoning about termination behavior. On the basis of this, we investigate the termination behavior of probabilistic programs. We introduce the notions of termination degree, bounded termination and path stoppability and investigate their mutual relationships. Path stoppability characterizes a broadened class of termination and allows for the computation of program runs that are otherwise considered as non-terminating.

Furthermore, we elaborate a denotational semantics for the probabilistic lambda calculus. The domains of this denotational semantics are probabilistic pre-distributions as base domains and ω-continuous function spaces as higher-type domains. We show the basic correspondence between the denotational semantics and the established Markov chain semantics.

Tallinn, November 2016 *Dirk Draheim*

Contents

1

Introduction

In this book, we are interested in systems that consist of programmable components and encounter probabilistic impact. We find such systems in many application areas, i.e., whenever a software-intensive system operates in a dynamic, vague environment: control systems, production systems, logistics systems, socio-technical systems of any kind. Also, the components of these systems themselves may show probabilistic behavior. However, probabilistic programs are interesting also in their own right, i.e., even if the probabilism is not a circumstance that we need to deal with, but is generated for the sake of probabilistic programming itself. This is the field of randomized algorithms. Randomized algorithms become ever more important in practice, in particular, in the field of cryptographic systems. In this book, we are interested in the semantics of probabilistic programs. And we are interested in systematic reasoning about probabilistic programs.

We take a foundational, completely reductionist approach. We narrow our investigation to a maximally reductionist programming language, the typed lambda calculus with recursion. We enrich this lambda calculus by a single programming construct for probabilistic choice, which yields the probabilistic typed lambda calculus, which we also often call just the probabilistic lambda calculus for short. We investigate both the operational semantics and the denotational semantics of the resulting calculus. First, we will delve into the operational semantics. On top of its basic operational semantics, we systematically give a Markov chain semantics to the probabilistic lambda calculus. This way we unlock the whole machinery of probability theory, in general, and Markov chains, in particular, for reasoning about probabilistic program systems. We use this mathematical machinery to systematically investigate the termination behavior of probabilistic programming systems. We will come up with a broadened notion of termination, so-called path stoppability. Path stoppable programs have a finite term cover, therefore, linear algebra can be exploited to determine the probability with which program outcomes are reached. Our investigation yields a precise infinitesimal understanding of the termination degree of a program.

Next, we will define a denotational semantics of the probabilistic lambda calculus. Denotational semantics is the Scott-Strachey approach to the semantics of programming languages [243]. The mathematical beauty of denotational semantics stems from its compositionality resp. de-compositionality. A denotational semantics is given inductively along the abstract syntax of a programming language. It establishes a correspondence of syntactical and semantical constructors. This way, a denotational semantics achieves implementation independency [238], i.e., it can be considered the specification of a programming language. Denotational semantics has developed into the de facto standard for the investigation of programming language semantics, see [192, 253, 224, 117, 103, 118]. Moreover, denotational semantics gave rise to domain theory, see [239, 240, 105, 116, 106] and also [224, 117, 3]. Domain theory provides and investigates appropriate mathematical structures for programming language semantics. However, domain theory is not only important when teamed together with denotational semantics, rather, it yielded important results in its own right. Most importantly, in [240] Dana Scott found a model for the untyped lambda calculus. In [242], Dana Scott incorporates probabilism into such a model. This way, the semantics for untyped probabilistic lambda calculi, called stochastic lambda calculi in [242], has been achieved.

In this book we deal with the typed probabilistic lambda calculus. And also with respect to denotational semantics we only deal with the typed version of the probabilistic lambda calculus. We will define a denotational semantics for the probabilistic lambda calculus based on ω-cpos (ω-complete partial orders). We use pure probability pre-distributions as ω-cpos. In the case of call-by-name semantics the probabilistic choices at higher types can be flattened denotationally to probabilistic choices of ground type. This way, in case of call-by-name we need distributions only in the construction of the base domains but no nested constructions as higher types. Also, we prove the basic corresponce between the denotational semantics and our operational Markov chain semantics.

In the upcoming sections we will give a more detailed overview of the motivation and the contributions as well as pointers to important literature. In Sect. 1.1 we motivate the book's investigations from the perspective of system simulation and system analytics as well as from the perspective of randomized algorithms. In Sect. 1.2 we outline the probabilistic lambda calculus and its Markov chain semantics. In Sect. 1.3 we explain what will be achieved with respect to the analysis of termination behavior of probabilistic programs. In Sect. 1.4 we give an outline of our denotational semantics. In Sect. 1.5 we give further remarks on the book's content and provide a chapter outline of the book.

1.1 Motivation

In the IT sector, we have many systems that are mere data-processing systems. Their task is to capture some data, maybe transactional, and to process and keep them for us [83, 82, 34, 89, 86, 79, 78]. We may find such systems in enterprise computing [85, 84, 75, 10], e.g., enterprise resource-planning systems, customer relationships management systems, any kind of master data management systems, e.g., identity management systems, any kind of data repositories, optical archives, information bases and so forth. Of course, all of these systems are no silos and offer interfaces to their environments. They are used by humans and also other information systems. However, the interaction with such systems is usually confined to data access.

On the other hand, we see many software-intensive systems that are highly engaged, i.e., actively engaged, with their environment. They react to external triggers. They adapt to the changing state of their environment. This is, in a sense, the realm of agent-oriented systems [261, 132, 133]. Examples stem from the field of control systems, robotics, manufacturing execution systems, production planning systems or logistics systems. The degree of interaction, e.g., in terms of frequency, reaction time or criticality may greatly vary. Also, the frontiers to the data-processing systems mentioned above vanish more and more in today's system landscapes: business process management systems [74, 7, 10, 73] become adaptive [167, 229, 260, 76, 135, 88, 8, 80], decision support systems become reactive [120, 256], etc. In general, IT systems become ever more adaptive. With cloud computing [100, 77], elasticity of IT infrastructure has become mainstream [188].

Also, we might want to deal with internal components of the system that show probabilistic behavior, i.e., we need to deal not only with external but also internal probabilistic events.

The described systems are all highly relevant and there are two very important and huge communities that deal with them, i.e., the community of *model checking* and the community of *model-based design*. We will give some further remarks on model checking and model-based design together with hints to the respective literature in due course in Sect. 1.5. On the programming side, reaction to probabilistic events always gives rise to non-determinism. If we know the statistical distribution of the relevant external events this opens the opportunity of probabilistic reasoning about the overall system. If we have a rigorous model of how probabilities propagate through the programmed systems, we can make assumptions on program outcomes and overall system behavior. Given the Six Sigma approach [125, 126] the potential of such statistically founded reasoning should be immediately clear for the field of numerical control and manufacturing execution systems. Consider an example from a higher-level use case, i.e., from the domain of logistics. Consider a stock management system. A stock management system would react to events concerning the amount of incoming goods, overrunning or too scarce stockpiles and maybe other variable resources such as availability of employees and so forth. Based on that it would automatically order goods or re-distribute goods. Now, as a mature option, the dataflow through a single storage could be modeled with queueing theory [101, 115, 268, 31], also, the supply chain consisting of a network of several storages could be modeled as a net of queues. Other options to model the storages and their network exist. For example, we could try to exploit stochastic extensions of Petri nets [119, 64, 13] or stochastic extensions of process algebra [107, 134, 50] for this endeavor. Anyhow, our reasoning about the programmed system in the single stores should fit into the overall probabilistic model to enable seamless reasoning. Therefore, we need a rigorous semantics of the programmed system involving probabilities. Our approach is reductionist. We will choose the lambda calculus as a candidate for our investigations. We will extend it by a probabilistic choice and give a Markov chain semantics to it. A probabilistic choice can be thought of as an input channel, yielding a reductionist, binary external information "left" or "right", i.e., the decision where to move next. This approach can be considered a first, fundamental step in the direction of an overall target – the switch from simulation of systems to a systematic analytical approach.

By the way, against the background of the above storage example, it is worth noting that there exist several business-process-modeling tools that offer features for process simulation. Simulation is exactly about forecasting the behavior of the modeled system like the flow of goods through a net of storages as described above. However, to our best knowledge, we do not know of a single business-process-modeling tool that incorporates queueing theory to support an analytical approach. This means, although with queueing theory we have a powerful tool to analyze systems, queueing theory is not yet consumable, i.e., it is not yet brought to the end-user, not yet brought to the average domain expert working in the field. Now, in computer science, or, to be more precise, in the field of software engineering, we actually have a long tradition in approaching simulation systematically. The original motivation of the first object-oriented programming language Simula [212] was to create a framework and language for system simulation. So, approaching simulation systematically was at the root of the early object-oriented paradigm. Only later, object-oriented programming turned into a paradigm of reusable [104], self-responsible software components that was particularly well-suited to serve the needs of upcoming object-oriented user interfaces. The original motivation towards system simulation has been a little bit forgotten in the object-oriented programming languages and design community. Instead, we can recognize the simulation approach as a basis of the agent-oriented system and modeling paradigm [261]. Nevertheless, in a sense, the original vision of Simula simply has been turned into reality. Today's object-oriented programming languages are mature candidates to specify executable system models – it's just about programming. What we need is to raise the level to system analytics. And there is a need to do this systematically, by building analytical features into existing programming and modeling environments. Even better, we should think about systematically applying automatic or semi-automatic reasoning platforms [40, 39, 209] in the field of software-intensive systems.

So far, we have considered probabilistic programming languages as embedded into probabilistic environments. Here, the probabilistic programming language is used to program components that have to react to probabilistic events. In this perspective, our reductionist probabilistic choice models an input channel that delivers information about events from the environment to the programmed component. However, probabilistic programming is important in its own right, in order to program algorithms stand-alone, without relationship to an environment. Here, the probabilistic choice is fed by a random generator, i.e., it is a programming element instead of representing an information channel to an outside system. This is the field of randomized algorithms. Randomized algorithms can speed up the solution of problems dramatically, at the price of yielding an erroneous result occasionally. For a seminal work on randomized algorithms see [226] by Michael O. Rabin, see also the work of Andrew C. Yao in [264] as well as Andrew C. Yao and F. Frances Yao in [265, 266]. The classical example of randomized algorithms is about primality testing, approached by Solovay and Strassen in [248] and Michael O. Rabin in [227]. For an overview on the topic of randomized algorithms see [205] and also [148]. For a thorough treatment of randomized algorithms including many use cases, see [204]. An introduction to the complexity theory of randomized algorithms can be found in the standard text book on automata theory [137].

Now, a formal semantics of probabilistic programming languages also allows for reasoning about randomized algorithms. Reductionist models of randomized algorithms have been given as probabilistic automata in [68] and [225]. For a formal treatment of complexity of randomized algorithms see the seminal text book on automata and language theory by Hopcroft, Motwani and Ullman [137], which encompasses the definition of complexity classes on the basis of the probabilistic Turing machine. An important source concerning the complexity theory of randomized algorithms can be found in the quantum computing literature. See [208] for a comprehensive text book on quantum computing and [25] for a survey of quantum computing complexity theory that also clarifies the relationship between the complexity of randomized algorithms and the complexity of quantum computing algorithms.

1.2 The Probabilistic Lambda Calculus

The typed lambda calculus with recursion can be considered a most reductionist functional programming language. It can be considered the essence of functional programming languages like ML [190, 111, 193] and Haskell [138], compare also with [24, 23, 102]. Furthermore, the typed lambda calculus with recursion has been subject to intensive investigation in the semantics of programming language community, where it is also called PCF (Programming

Language for Computable Functions) [223] on many occasions. Therefore, we have chosen it as our candidate for investigating the semantics of probabilistic computation. It is the task of this section to explain how the probabilistic lambda calculus emerges as an extension of the typed lambda calculus with recursion, see Figs. 1.1 and 1.2. We have preferred the probabilistic lambda calculus over other options such as the probabilistic Turing machine [68, 225]. As opposed to the Turing machine, the lambda calculus comes with a particular 3GL (third generation language) or even 4GL (fourth-generation language) look and feel, albeit in a most reductionist form. Programs of the probabilistic lambda calculus are particularly intuitive and easy to read, because of their high similarity to the mathematical notation of recursive functions. Similarly, the probabilistic lambda calculus is particularly amenable to a denotational treatment that we will also provide in this book.

1.2.1 The Typed Lambda Calculus with Recursion

The lambda calculus in its original form, as introduced by Alonzo Church [47, 48], is an untyped language. If the lambda calculus is enriched by a type system, it is also called the simply typed lambda calculus [19]. In order to gain a Turing-complete typed calculus, an explicit recursion construct μ is added to the simply typed lambda calculus. Then, the resulting calculus is usually called the typed lambda calculus with recursion. Henceforth, we will also refer to this calculus simply as the lambda calculus or λ-calculus if it is clear from the context which calculus we mean.

Basically, the lambda calculus is introduced as a syntax and its operational semantics – see Fig. 1.1. The syntax is defined as context-free syntax plus a type system which specifies the notion of well-typed term. The operational semantics is introduced in two stages. First, the so-called immediate reduction relation is defined. This specifies which single steps are possible between terms. Actually, the immediate reduction relation is a partial function from terms to terms. The constants are considered the result values of programs. The constants and all of those terms that are outermost abstractions are the so-called values of the lambda calculus. It is not possible to do a further step from a value. Values stop computations. For each other term M there exists exactly one successor term N to which a next step is possible, which is denoted by $M \to N$. This right-uniqueness of the immediate reduction relation makes the considered lambda calculus a deterministic calculus. The considered lambda calculus is deterministic, because a concrete reduction strategy, in our case call-by-name, is fixed for it.

The immediate reduction relation forms the first stage of the lambda calculus' operational semantics. As the second stage the reduction relation is defined as the transitive, reflexive closure of the immediate reduction relation. We say that a term M reduces to another term N if it is connected to it – in the correct direction – via the reduction relation, which is denoted by

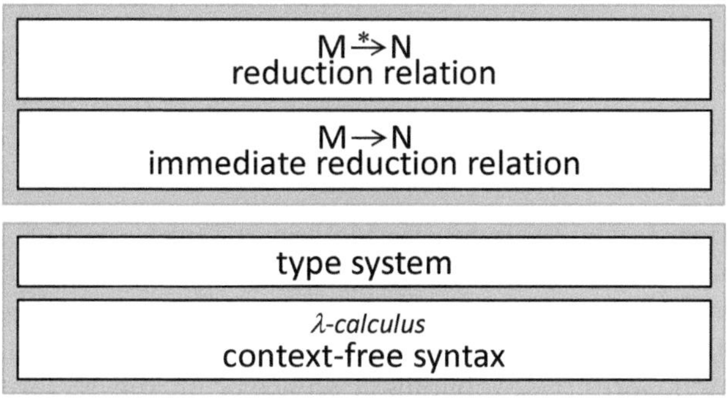

Fig. 1.1. Syntax and semantics of the λ-calculus

$M \xrightarrow{\star} N$. We then also say that N is reachable from M. It is very interesting to recognize that the specification of the immediate reduction relation is already sufficient for an implementation of the lambda calculus as a programming language. It might be a bit unfair, but we could say that the reduction relation does not add anything else but some terminology, i.e., the notion of reachability of terms. It is actually unfair, because semantics is, first, about the agreement of what we actually intend with a formal language and, second, a means to get access to further semantical tools that can be used in reasoning about the program. Obviously, we need to agree upon the notion of program results and, furthermore, an understanding of the reduction relation as a transitive closure will be needed in formal argumentations on program behavior. Nevertheless, the definition of the immediate reduction relation is not sufficient for all investigations of program behavior that we might want to conduct. Further notions, such as program reduction trees, are usually defined on top of it to gain appropriate mathematical tools for program reasoning. Anyhow, it is worth noting that with respect to the behavior of programs, the immediate reduction relation is already a complete specification. This viewpoint will also be useful in the comparison with the probabilistic lambda calculus in due course.

In general, it is not necessary to fix a reduction strategy for a lambda calculus, so that a lambda calculus may encounter some level of non-determinism. However, such non-determinism does no harm, or, to say it better, makes no difference with respect to the program outcomes. This is so due to the Church-Rosser property [49]. If a program is able to reduce to a constant, i.e., if it is a terminating program, then this constant is uniquely given. Furthermore, we can reduce to it from any reachable intermediate term. However, the latter is not so important for us here. What interests us here is the uniqueness of terminating program outcomes. This actually guarantees a certain level of determinism. With respect to terminating program results, each lambda calculus is deterministic, independent of the chosen reduction strategy. And here also lies the difference between the non-determinism encountered in lambda calculi in general as opposed to the probabilistic lambda calculus in this book. In the probabilistic lambda calculus, a program may terminate with different constants.

1.2.2 The Probabilistic Lambda Calculus Compared

Now, let us turn to the probabilistic lambda calculus, see Fig. 1.2. Syntactically, the probabilistic lambda calculus is just the typed lambda calculus with recursion plus a program construct for probabilistic choice. Of course also the type system is adjusted to the new terms. Now, for any two terms M and N, the term $M|N$ denotes the probabilistic choice of M and N. Given a program $M|N$, this program executes with fifty per cent probability as M and, equally, with fifty per cent as N. Again, we give the semantics in two stages. First we define a one-step semantics. However, this time the one-step semantics is a total function that assigns to each pair of terms M and N the probability i with which the program may move from M to N, which is then denoted by $M \xrightarrow{i} N$.

Of course, all terms of the plain lambda calculus are also terms of the probabilistic lambda calculus. Each immediate reduction $M \rightarrow N$ can be found in the probabilistic lambda calculus as step $M \xrightarrow{1} N$. Of course, the crucial difference lies in terms of the form $M|N$. In case that $M \neq N$ we specify two possible reductions, i.e., a reduction $M|N \xrightarrow{0.5} M$ and a reduction $M|N \xrightarrow{0.5} N$. For each term P that is different from both M and N, we define an immediate one step reduction $M|N \xrightarrow{0} P$ to ensure that the one-step reduction becomes a total function. In case of terms $M|M$ we will introduce a possible reduction $M|M \xrightarrow{1} M$. Again, for each term P different from M we define a one-step reduction $M|M \xrightarrow{0} P$.

On top of the one-step semantics we then define the Markov chain semantics of the probabilistic lambda calculus, see Fig. 1.2. We take closed terms of the probabilistic lambda calculus as the states of a Markov chain S. We take the one-step semantics as the transition matrix of this Markov chain. Now, we define the reduction semantics of the probabilistic lambda calculus via hitting probabilities of the Markov chain. The probability to reduce from a term M to a term N, denoted by $M \Rightarrow N$, is defined as the Markov chain's probability of starting in M and ever hitting N. With the Markov chain semantics we inherit all results from probability theory and Markov chains for reasoning about program behavior of probabilistic programs. For example, we can determine reduction probabilities as least solutions of linear equation systems.

In this book we consider the lambda calculus under the call-by-name strategy. This is a very common choice: note that also PCF [223] is considered with the call-by-name strategy. The choice of call-by-name is convenient. Actually, the results on termination behavior in Chap. 4 are independent of the call-by-name strategy, i.e., they would also hold under a concrete call-by-value

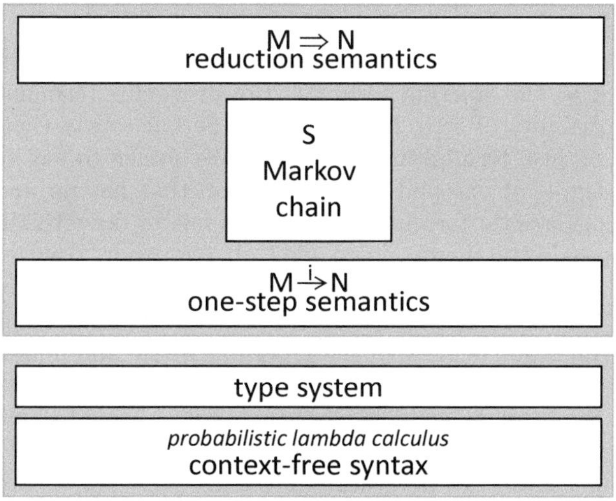

Fig. 1.2. Syntax and semantics of the probabilistic lambda calculus

strategy. However, the choice of call-by-name is crucial for the denotational semantics in Chap. 5. Here, the choice of call-by-value would result in different domains than the chosen ones.

1.3 Termination Behavior of Probabilistic Programs

In the deterministic lambda calculus a program either terminates or does not. Each program corresponds to exactly one program run. This is not so any more with the probabilistic lambda calculus. Here, a program results, in general, in one out of many possible program runs. It is the single program run that may terminate or not. Now, a non-deterministic program may have some terminating program runs plus some non-terminating program runs. However, we will define the notion of termination degree for probabilistic programs. The termination degree of a program is the probability that it will ever reach one of the constant values. Or, to say it differently, the termination degree of a program is the accumulated probability of all of its terminating program runs.

Let us coin the somehow artificial term of strictly terminating program. We say that a program strictly terminates if all of its program runs terminate. Actually, with respect to terminology we are in a bit of a dilemma. We would like to avoid calling a non-deterministic program a terminating program. The problem is that we have the interesting class of programs with termination degree one. In this class, there are also programs that may have a non-terminating program run. For example, the program $M = \mu\lambda x.(x|0)$ has a termination degree of one, however, it also has a non-terminating program run. The program M will reach the constant 0 with a hundred per cent probability. On the other hand M has a non-terminating program run $M \xrightarrow{1} \lambda x.(x|0)\mu\lambda x.(x|0) \xrightarrow{1} (M|x) \xrightarrow{0.5} M \xrightarrow{1} \cdots$ which is kept in an endless loop back to the starting term M. However, this terminating program run has a probability of zero. Now, it would be fair to say that M is always terminating, or just terminating for short. We prefer to say that M has a termination degree of one and that a program that has no non-terminating program runs is strictly terminating. All this has to do with the original *explicatio* of probability theory given by Kolmogorov in [163] as a model of experimental data, and with how we usually speak about events that have zero per cent probabilities as impossible events. The termination degree of M equals the termination degrees of the programs $\mu\lambda x.0$ and 0 in our semantics, i.e., they are *infinitesimally* "equal".

Based on the Markov chain semantics we will be able to identify a broadened notion of termination, so-called path stoppability, see Fig. 1.3. Path stoppability is a notion of program analysis. It allows us to stop some non-terminating programs and to determine their termination degree. Let us walk through the example programs given in Fig. 1.3. The program $(\lambda x.x)(0|1)$ is a strictly terminating program. All of its program runs terminate. This is not so for the program $M = \mu\lambda x.(x|0)$. We have just discussed that this program has a non-terminating program run. We will design an algorithm, let us call it the path-stopping algorithm, that dovetails a given non-deterministic program and detects and stops all of its endlessly looping program runs. We will show that this algorithm terminates for all of those programs that have a finite cover. The path-stopping algorithm implements a program analysis. Now, program M has a finite cover, and therefore it is stoppable by the path-stopping algorithm. It falls into the class of path stoppable programs. This is not so for the program $(\mu\lambda x.(+1(x) \mid 0))$. The cover of this program contains infinitely many terms, e.g., all terms of the form $+1^n(0)$ for each number n.

The notion of path stoppability also applies to deterministic lambda programs. We have given the corresponding examples in Fig. 1.3. The program $(\lambda x.x)0$ serves as an example for strictly terminating programs, the most simple non-terminating program $\mu\lambda x.x$ as an example for path stoppable programs and the program $\mu\lambda x. + 1(x)$ as an example for programs that are not path stoppable.

For path stoppable programs, it is possible to compute their termination degrees. As a consequence, it is possible to compute all reduction probabilities for path stoppable programs. We will also speak of path computability or p-computability of reduction probabilities. In order to investigate the termination behavior of probabilistic programs as just outlined, we will need results from graph theory. Therefore, the Markov chain semantics is teamed together with the notion of reduction graph and the notion of reduction tree – see Fig. 1.4. We will interpret the one-step semantics as a graph, the so-called reduction graph R. Based on that we will precisely define program executions, program runs and term covers. We will tightly integrate the graph seman-

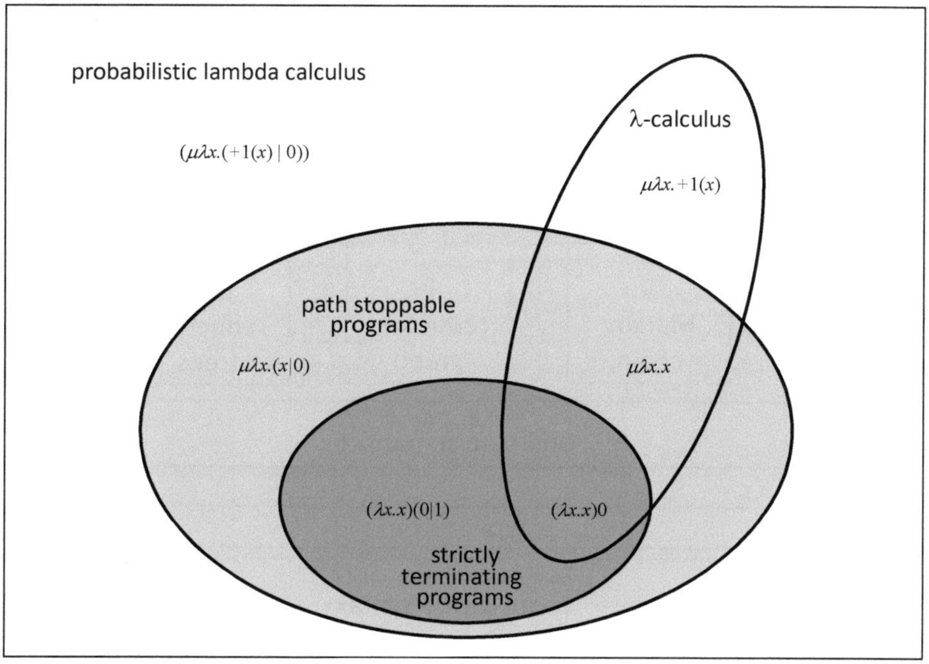

Fig. 1.3. The λ-calculus and the probabilistic lambda calculus compared

tics with the Markov chain semantics. This way we unlock graph theoretical results for reasoning about programs and their behavior. For example, we will exploit König's lemma in the investigation of termination behavior of programs. Similarly, we define a tree semantics on top of the one-step semantics. For each program M we define the tree $\tau[M]$ of program runs starting in M. Reduction trees are defined on the basis of the reduction graph R. They provide a tree-specific viewpoint on the reduction graph and unlock tree-specific graph theoretical results for program reasoning. For example we will use Beth's Tree Theorem, which is an instance of König's Lemma, in the investigation of bounded program termination.

1.4 Denotational Semantics

Denotational semantics is the Scott-Strachey approach to the semantics of programming languages [243]. The denotational semantics of a programming language is given directly in terms of the mathematical objects that are computed by the programs of a programming language. A denotational semantics targets implementation independency [238], i.e., it can be considered the specification of a programming language as opposed to the several possible implementations of this programming language. A major characteristic of denotational semantics is compositionality. This means that a denotational semantics

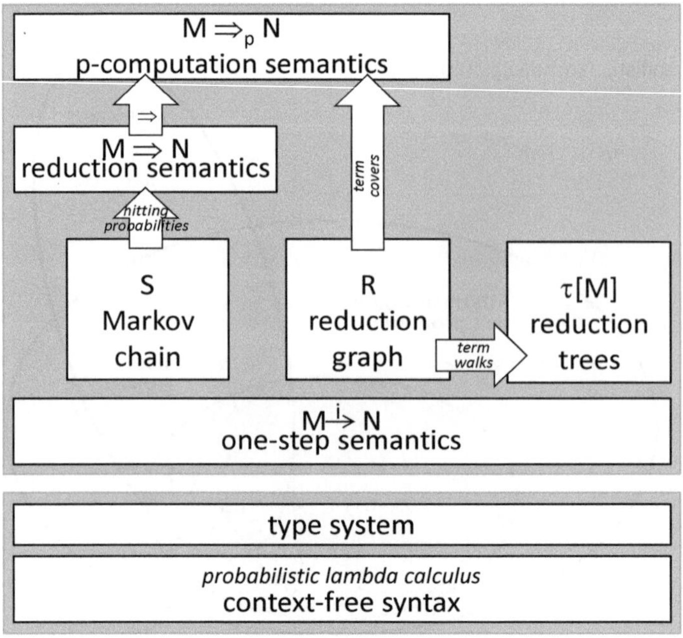

Fig. 1.4. Auxiliary operational concepts for the probabilistic lambda calculus

introduces an appropriate semantic constructor for each of the syntactic constructors of the programming language. Then, the denotational semantics is given inductively along the abstract syntax of the programming language and the correspondence of syntactical and semantical constructors. Given this inductiveness, a denotational semantics also reveals operational semantics, i.e., it can be considered a program of a recursive "meta" programming language. The point is that it does not have to rely on an operational semantics of this pre-assumed meta programming language, it itself receives its semantics from first principles, i.e., the notion of inductive definitions. For more on inductive definitions see also our primer on this topic in this book in Sect. 2.4.

Against the background of implementation independence, a denotational semantics can be considered as coming first. Then, it can be considered a specification in a programming language engineering process. However, also the opposite perspective is admissible. Here, the denotational semantics is defined for an existing programming language for the purpose of clarifying its semantics and, even more important, for unlocking the mathematical toolkit for reasoning about program behavior. At least when we treat reductionist programming languages like the lambda calculus, denotational semantics sometimes rather has this flavor of coming second, after the operational semantics.

The reductionist programming language and reasoning system Edinburgh LCF (Logics for Computable Functions) [237, 190, 191, 111], also just LCF for short, has been given a denotational semantics in [192]. The reductionist programming language PCF, which is also based on LCF, has been given a denotational semantics in [223]. A major task in denotational semantics is to establish the appropriate domains from which the semantical objects are drawn. A major challenge is to provide domains for recursively defined data types which naturally arise in programming languages. Dana Scott showed how to solve this problem of finding solutions of recursive domain equations even if function spaces are involved. In [240] he was able to construct a model of the untyped lambda calculus. A model of the untyped lambda calculus amounts to a function space that is isomorphic to itself, i.e., is a solution to the domain equation $D \cong D \to D$. In general, i.e., in the case of sets, it is impossible to find a solution to the domain equation $D \cong D \to D$, because the cardinality of $D \to D$ is larger than the cardinality of D for any set D. In [240] such a function space has been established on the basis of complete lattices.

The construction of domains for denotational semantics evolved into a discipline in its own right, i.e., domain theory. The foundational challenge addressed by domain theory is to tighten the correspondence between the denotational semantics and the operational semantics of programming languages and to provide optimized domains for this purpose. Orthogonal to this challenge, domain theory provides specialized domains to address various programming language phenomena such as non-determinism or parallelism. For texts on domain theory, see [116, 224, 3]. For comprehensive texts on denotational semantics in general, see, e.g., [253, 235, 117, 203, 118, 263].

We will elaborate a denotational semantics for the probabilistic lambda calculus. The chosen approach is straightforward and standard from the literature; compare with the work of Saheb-Djaromi in [232, 233]. Our denotational semantics is based on ω-complete partial orders, i.e., ω-cpo*s*. We have that ω-cpo*s* are, among other domains such as, e.g., complete lattices, directed complete partial orders, a well-known choice for denotational semantics. They have been used very early for this purpose, i.e., in the definition of a denotational semantics of Edinburgh LCF, see [192]. As basic mathematical objects of our denotational semantics, we will use vectors of real numbers in $[0,1]$. These vectors assign probability values to data points. We call these vector probability pre-distributions, because they play the same role as distributions. There is a crucial difference to distributions. Their total mass, i.e., the sum of all the values of all possible data points, does not necessarily have to be a hundred per cent in case of pre-distributions. And this makes sense. A total mass of less than a hundred per cent stands for the existence of some non-terminating program runs. Actually, we will call the total mass of a pre-distribution the degree of termination later. As a result, we work without explicit bottom elements representing non-termination. Bottom elements arise implicitly as distributions that assign a zero per cent probability to each data point. Furthermore, we will see that probabilistic choices at higher types can be flattened completely to probabilistic choices of ground type. This way, we need distributions only in the construction of the base domains. All the domains of higher type that we introduce in our semantics turn out to be vector spaces, which greatly eases our argumentations and formal proofs.

Given a program M of the probabilistic lambda calculus, we will denote, as usual, the semantical object assigned to it as $[\![M]\!]$. Given a number constant n_i, its semantics is given as the (pre-)distribution that assigns a hundred per cent probability to the so-called data point i, i.e., $[\![n_i]\!](i) = 1$ and $[\![n_i]\!](j) = 0$ for all $j \neq i$. Next, we also say that n_i represents the data point i and use $[n_i]$ to denote it, i.e., $[n_i] = i$. Similar definitions can be given for the other ground type of the probabilistic lambda calculus, i.e., Boolean. Once we have introduced the denotational semantics for the probabilistic lambda calculus, we will investigate its relationship to the operational semantics. We will prove the one-to-one correspondence at the base element level. More concretely, this correspondence means that the semantics of a program M, applied to a data point $[c]$, equals the probability with which M reduces to that constant c that represents $[c]$, which can expressed in the notation that we have introduced so far as follows: $([\![M]\!][c]) = (M \Rightarrow c)$.

1.5 Chapter Outline and Further Remarks

Chapter 2 is a preparatory chapter. It recaps and comments on basic mathematical tools needed throughout the book, in particular, from the fields of probability theory, Markov chains, graph theory and domain theory. Also, it delves into the topic of inductive definitions, because they form the foundation of rigorous specification of semantics. In Chap. 3 we define the syntax and establish the Markov chain semantics of the probabilistic lambda calculus. To improve comparability we also introduce the operational semantics of the plain, deterministic lambda calculus. In Chap. 4 we investigate the termination behavior of the probabilistic lambda calculus. We define the notion of termination degree. Furthermore, we define the notion of bounded termination, which is about programs that do not exceed an upper bound of steps whenever they terminate. We define the notion of path stoppability that we have described above as a broadened notion of termination. Then, we systematically investigate the mutual dependencies between path stoppability, bounded termination and termination degrees of a hundred per cent. To achieve all this, the chapter establishes the graph semantics as well as the tree semantics of the probabilistic lambda calculus. Also, it shows some needed graph cover lemmas, i.e., the fact that the cover of a graph is already completely determined by the cover of its paths and, second, that the finite cover of a graph is computable for all k-ary graphs that have a finite cover and a computable edge relation. Finally, in Chap. 5 we define a denotational semantics of the probabilistic lambda calculus, based on functionals over probability distributions as domains. As the basic semantic correspondence, we show the correspondence of the denotational semantics with respect to the Markov chain semantics.

The main focus of this book is the Markov chain semantics for the probabilistic higher-order typed lambda calculus. It is a natural idea to treat probabilistic computation with Markov chains, compare, e.g., to the textbook [204] of Motwani and Raghavan on randomized algorithms and, again, to the work on the probabilistic lambda calculus by Saheb-Djahromi in [232, 233]. Compare also with, e.g., [128, 174, 38, 94, 110, 175, 72] to name a few. The book's aim is to fully elaborate the Markov chain semantics, i.e., to elaborate it to a level that systematically unlocks results from Markov chain theory to be used in reasoning about program semantics and, in particular, establishing further formal results. This can be understood best by looking at how we exploit the established semantics to investigate termination behavior of probabilistic programs and come up with and further investigate notions like path stoppability, path computability and bounded termination.

The focus of the book is narrow in several ways, i.e., with respect to the probabilistic programming phenomena it delves into, the chosen programming language paradigm, the programming language semantic approaches it exploits, the level of investigation and the motivation it stresses for its investigations. All of these aspects are mutually dependent. First, we deal with the functional programming language paradigm and here we have chosen the most reductionist language, i.e., the typed lambda calculus also known as Plotkin's PCF [223]. A thorough treatment of probabilistic imperative programming is provided by the book [185] by McIver and Morgan.

We delve into operational semantics. Also, we work with denotational semantics. We treat denotational semantics only as far as we feel is needed to bridge our treatment into the extremely mature body of knowledge that is established by denotational semantics and domain theory for probabilistic computation, see Sect. 5.4 on selected important readings in the field. We do not look into axiomatic semantics [136] of probabilistic computing. Once more, we want to recommend the book [185] by McIver and Morgan as an authoritative reference. The important field of axiomatic semantics is so rich with respect to probabilistic computation, we do not even attempt to give a literature overview, again, the reader might want to use [185] as a good starting point.

The field of model checking is extremely mature with respect to probabilistic systems. The overview article [175] of Legay, Delahaye and Bensalem provides a good entry point into the subject matter. The reason why the field of model checking is particularly important is because it does not stop at defining and investigating formal logics and reasoning systems, but actually provides concrete implementations of model-checking platforms or model checkers. Important probabilistic model checkers are ETMCC [131], Prism [172], Ymer [267], Vesta [244], and MRMC [149]. The list does not aim to be complete, nor does it express preference or priority. A rigorous comparison of these model checkers is provided in [140] by Jansen et al. A typical model checker for probabilistic systems allows us to model systems as discrete- and continuous Markov chains. Based on these models, it allows for system simulation and automated or semi-automated verification of system properties. Probabilistic versions of the temporal logics CTL and CTL*, i.e., pCTL and pCTL*, have been introduced by Hansson and Jonsson in [121]. In [11, 12], Aziz et al. introduce the logic CSL (Continuous Stochastic Logic), which allows for reasoning about continuous-time Markov chains, compare also with [14], which treats, in more depth, model checking aspects of CSL. As we said, we consider model checking based on Markov chain models as typical, albeit other important approaches exist such as, e.g., the application [30] of uniform continuous-time Markov decision processes [15] to statecharts [122, 123], the GreatSPN tool [64, 13] with respect to Petri nets, or the PEPA workbench [107, 134] with respect to stochastic process algebras.

It is also important to mention the industrial-strength modeling environment Simulink [257] and, in particular, also its module SimEvents [114, 44]. Simulink is based on Matlab. It supports model-based design, a combination of modeling, simulation, code generation for and verification of dynamic systems consisting of both discrete and continuous switching blocks. Its original domain is the domain of control systems and it is applied at several levels and instances thereof ranging from manufacturing execution systems over embedded systems to circuit design. The module SimEvents extends Simulink by queueing-system building blocks.

Programming languages such as IBAL [218], Church [110] and Venture [180] are programming languages that contain a probabilistic programming primitive. The purpose of these languages is to express stochastic models, i.e., to generate stochastic models. They allow for querying distribution outcomes against the traces of a probabilistic program. This way, these programming languages become decision support tools. They are called probabilistic programming languages by Stuart Russell in [231] or stochastic programming languages in [110], however, the language IBAL is called a rational programming language by Pfeffer in [218], because the purpose of such languages is to support reasoning about rational agents. A precursor of IBAL was already developed by Pfeffer et al. in [161].

2

Preliminary Mathematics

This chapter provides basic mathematical definitions and propositions needed in this book. The chapter is a preparatory chapter that can be skipped on first reading without major loss. Nevertheless, the chapter sets the stage for the upcoming material and it might be a good means to tune in, so I recommend at least glancing at the material before turning to the real content of the book. Note that I have tried to motivate each piece of mathematics in this chapter, i.e., I would like to make clear why it is important and where it is used in the book. I tried to create a living repository rather than a dead list of definitions and theorems. Of course, a sweet spot had to be maintained between readability and conciseness. The purpose of this chapter is many-fold. In the first place, it should serve as a quick and easy reference for the study of the upcoming material. In doing so, the chapter also aims at streamlining the notation used in the book. Last, but not least, we give pointers to the literature, both to foundational texts as well as to further readings. However, the major objective of the resulting book organization was to extract necessary standard material from the content chapters into its own, dedicated chapter in order to create lean and better readable content chapters. We only provide proofs in this chapter whenever we think that the proof adds to the understanding of other parts of the book.

We start with basic material from probability theory in Sect. 2.1, which is at the core of the book's topic. Then, we present basic material in Markov chains in Sect. 2.2. In the book we will exploit Markov chains to give systematic operational semantics to probabilistic computations. Next, in Sect. 2.3 we turn to graph theory which will be needed as a tool to study and analyze the operational semantics of the probabilistic lambda calculus, in particular, to establish results about termination behavior. Then in Sect. 2.4 we discuss various means and approaches of inductive definitions. Inductive definitions provide the foundations of denotational semantics. Actually, at an appropriate conceptual level, the discipline of denotational semantics can be identified with the field of inductive definitions. Further mathematical material and tools are compiled in Sect. 2.5.

2.1 Probability Theory

Probability theory is the fundament of the book's material. The original work of Andrey Kolmogorov [162, 163, 164] is a very important source. See [37, 46, 251] for classic textbooks on the material and furthermore, e.g., [45, 17, 93, 22, 28, 95, 230]. See [163] on how probability spaces model experimental scenarios. For a rather Bayesian interpretation of probability theory, we recommend [141] as a mature discussion. The wide spread and classic frequentist interpretation of probability theory is completely sufficient for this book.

2.1.1 Probability Spaces

A probability space consists of a set of outcomes, a set of events and a probability function. Events are modeled as sets of outcomes. The set of events forms a sigma algebra over the set of outcomes, which means that it contains the whole set of outcomes, the empty set and is closed under complement and union of countable subsets. The probability function assigns real values as probability values to the events and fulfills the three axioms of Kolmogorov, i.e., the probability values are in the range from zero to one, the probability value of the whole set of outcomes equals one, and the probability of a countable set of pairwise disjoint events equals the sum of all of the individual events.

Definition 2.1 (σ-Algebra) *Given a set Ω, a σ-Algebra Σ over Ω is a set of subsets of Ω, i.e., $\Sigma \subseteq \mathbb{P}(\Omega)$, so that the following conditions hold true:*

1) $\Omega \in \Sigma$
2) *If* $\alpha \in \Sigma$ *then* $\Omega \backslash \alpha \in \Sigma$
3) *For all countable subsets of* Σ, *i.e., for all sequences* $\alpha_0 \in \Sigma, \alpha_1 \in \Sigma, \alpha_2 \in \Sigma, \ldots$ *it holds true that*

$$\bigcup_{i=0}^{\infty} \alpha_i \in \Sigma$$

Definition 2.2 (Probability Space) *A probability space $(\Omega, \Sigma, \mathsf{P})$ consists of a set of outcomes Ω, a σ-algebra of (random) events Σ over the set of outcomes Ω and a probability function $\mathsf{P} : \Sigma \rightarrow \mathbb{R}$, also called a probability measure, so that the following axioms hold true:*

1) $\forall \alpha \in \Sigma \,.\, 0 \leqslant \mathsf{P}(\alpha) \leqslant 1$ *(i.e.,* $\mathsf{P} : \Sigma \rightarrow [0,1]$*)*
2) $\mathsf{P}(\Omega) = 1$
3) *(Countable Additivity): For all countable sets of pairwise disjoint events, i.e., for all sequences* $A_0 \in \Sigma, A_1 \in \Sigma, A_2 \in \Sigma, \ldots$ *with* $A_i \cap A_j = \emptyset$ *for all* $i \neq j$, *it holds true that*

$$\mathsf{P}(\bigcup_{i=0}^{\infty} A_i) = \sum_{i=0}^{\infty} \mathsf{P}(A_i)$$

Note that it is a common to call the axioms (1) through (3) in Def. 2.2 the three axioms of Kolmogorov. The original formulation in [162, 163] consists of six axioms, i.e., (i) Σ is a field [129], (ii) Σ contains Ω, (iii) $\forall A \subseteq \Omega . \mathsf{P}(A) \geqslant 0$, (iv) $\mathsf{P}(\Omega) = 1$, (v) $A \cap B = \emptyset \implies \mathsf{P}(A + B) = \mathsf{P}(A) + \mathsf{P}(B)$ and (vi) the axiom of continuity, which requires

$$A_0 \supset A_1 \supset A_2 \supset \ldots \textit{and} \bigcap_{i \geqslant 0} A_i = \emptyset \textit{ implies } \lim_{n \to \infty} \mathsf{P}(A_n) = 0 \tag{2.1}$$

Then, in [162, 163], countable additivity, i.e., axiom (3) in 2.2 is proven equivalent to Eqn. (2.1) by the generalized addition theorem.

Often, we use $A \wedge B$ as well as A, B to denote the intersection $A \cap B$ of events A and B. Such notation is very common, in particular to denote probabilities of events, e.g., as in $\mathsf{P}(A \wedge B)$ or $\mathsf{P}(A_1, \ldots, A_n)$. We always use the notation that seems most convenient to us in the respective context. With respect to conditional probabilities, we equally use the notation $\mathsf{P}(A \mid B)$ as well as the notation $\mathsf{P}_B(A)$, depending on what is more telling or more concise in a given situation. Usually, we use the first form if the description of the event B is rather too complex to be used as an index, e.g., in the case of the Markov property in Eqn. (2.13).

Definition 2.3 (Conditional Probability) *Given events A and B, the probability $\mathsf{P}(A \mid B)$ of A conditional on B, also denoted by $\mathsf{P}_B(A)$, is defined as*

$$\mathsf{P}(A \mid B) = \frac{\mathsf{P}(A \wedge B)}{\mathsf{P}(B)} \tag{2.2}$$

Given a probability space $(\Omega, \Sigma, \mathsf{P})$ we call a countable collection of pairwise disjoint events $E \subseteq \Sigma$ a discrete probability distribution if the individual probabilities of the events in E sum up to one. Basically, in this book we are interested in discrete probability distributions and not in continuous probability distributions. Therefore we also talk about discrete probability distributions simply as probability distributions for short, or, even shorter, as distributions. Usually, a random variable, see Sect. 2.1.3, is used to characterize a distribution as the collection of inverse images under the random variable. Note that the collection of inverse images under a random variable is necessarily a collection of disjoint events, which is due to the right-uniqueness of each random variable. Now, it is common to call each function from an arbitrary countable index set I into the range of probability values $[0, 1]$ a probability function, as long as all values of the function sum up to one. We adopt this terminology in this book. In particular, we also need a notion of pre-distribution later, which is an approximation to a distribution. A probabilistic pre-distribution is then a function from a countable index set into the range $[0, 1]$, which sums up to a value less than or equal to one; compare with Def. 5.1. Furthermore, we call the sum of all values of a real-valued function the total mass of this function.

Definition 2.4 (Probability Distribution) *A function $I : S \longrightarrow [0,1]$ is called a discrete probability distribution, or distribution for short,* **iff** *its domain I is at most countably infinite and its total mass sums up to 1, i.e.,*

$$\sum_{s \in S} P(s) = 1 \tag{2.3}$$

For us, the original *explicatio* of probability by Kolmogorow, i.e., the *deduction* of probability theory from its primary application field, which is the observation of experimental data, is very important. Here [163], it is explicitly stated that an event that has probability zero is not impossible. Rather, it is said that an event with probability zero is practically impossible, however, the notion of practical impossibility has been created to have a concept that is different from impossibility. Furthermore, it is said that all the events of a probability space are considered *a priori* as possible. This means if we want to model a certain outcome as impossible, we should, from the outset, not include it in the set Ω of outcomes of the probability space. In particular, with our Markov chain semantics for the probabilistic lambda calculus we leave open the question whether a probability of zero in our transition matrix stands for an impossible or a practically impossible move. All of this should not bother us, it is not too important for the development of the subsequent theory, but should be kept in mind for further, advanced discussions. Either way, in this book we prefer to talk about events with a 100% probability or events with a zero percent probability instead of sure and impossible events.

2.1.2 Properties of Probability Spaces

In this section we list some basic properties of each probability space $(\Omega, \Sigma, \mathsf{P})$. The facts are an application of measure theory, i.e., the properties follow from the fact that (Ω, Σ) is a measurable space and $(\Omega, \Sigma, \mathsf{P})$ is a measure space with measure P, see, e.g., [17, 28, 22, 230], compare also with Def. 2.9.

Lemma 2.5 (Monotonicity) *If $A \subset B$ for $A, B \in \Sigma$ then $P(A) \leqslant P(B)$.*

Lemma 2.6 (Sub-Additivity) *Given a countable set $A_0, A_1, A_2, \ldots$ of events $A_i \in \Sigma$ we have that*

$$\mathsf{P}(\bigcup_{i=0}^{\infty} A_i) \leqslant \sum_{i=1}^{\infty} \mathsf{P}(A_i) \tag{2.4}$$

Lemma 2.7 (Continuity from Below) *Given a (strictly) ascending chain of events, i.e., a countable set $A_0 \subset A_1 \subset \cdots$ of events $A_i \in \Sigma$ we have that*

$$\mathsf{P}(\bigcup_{n=0}^{\infty} A_n) = \lim_{n \to \infty} \mathsf{P}(A_n) \tag{2.5}$$

Lemma 2.8 (Continuity from Above) *Given a (strictly) descending chain of events, i.e., a countable set $A_0 \supset A_1 \supset \cdots$ of events $A_i \in \Sigma$ we have that*

$$\mathsf{P}(\bigcap_{n=0}^{\infty} A_n) = \lim_{n\to\infty} \mathsf{P}(A_n) \tag{2.6}$$

In case the intersection of all A_n is empty in Eqn. (2.6) , we have that $\lim_{n\to\infty} \mathsf{P}(A_n)$ equals zero due to $\mathsf{P}(\emptyset) = 0$. Therefore, Lemma 2.8 is a generalization of the original sixth axiom of Kolmogorov [163], i.e., the axiom of continuity, see also the remarks on Eqn. (2.1) above.

2.1.3 Random Variables

Given a probability space, a random variable is a means to specify events and probability distributions. A random variable is a measurable function from the outcome space to an indicator set. The chosen indicator set and the specification of the function are used to model concrete probability distributions. Random variables form the basis of Markov chains, see Sect. 2.2. We will use random variables in the operational semantics of the probabilistic lambda calculus to model the fact that one term reduces to another term in n steps with a certain probability.

Definition 2.9 (Measurable Space, Measurable Function) *Given two measurable spaces (X, Σ) and (Y, Σ'), i.e., sets X and Y equipped with a σ-algebra Σ over X and a σ-algebra Σ' over Y. A function $f : X \to Y$ is called a measurable function, also written as $f : (X, \Sigma) \to (Y, \Sigma')$, if for all sets $U \in \Sigma'$ we have that the inverse image $f^{-1}(U)$ is an element of Σ.*

Definition 2.10 (Random Variable) *A random variable X based on a probability space $(\Omega, \Sigma, \mathsf{P})$ is a measurable function $X : (\Omega, \Sigma) \to (I, \Sigma')$ for a measurable space (I, Σ') with indicator set I. The notation $(X = i)$ is used to denote the inverse image $X^{-1}(i)$ of an element $i \in I$ under f, i.e.,*

$$(X = i) =_{DEF} X^{-1}(i) = \{\omega \in \Omega \mid X(\omega) = i\} \tag{2.7}$$

$$\mathsf{P}(X = i) =_{DEF} \mathsf{P}(X^{-1}(i)) = \mathsf{P}(\{\omega \in \Omega \mid X(\omega) = i\}) \tag{2.8}$$

Henceforth, we usually omit the σ-algebras in the definition of concrete random variables $X : (\Omega, \Sigma) \to (I, \Sigma')$ and specify them in terms of functions $X : \Omega \to I$ only. The notation $(X = i)$ is a typical notation used in the context of random variables. Given a random variable X, the expression $(X = i)$ is just an alternative means to denote the inverse image $X^{-1}(i)$ of i under the function X. The set $(X = i)$ is a subset of Ω and therefore denotes an event. The random variable notation for events can be extended in a natural, obvious way. For example, given a random variable $X : \Omega \longrightarrow \mathbb{N}_0$ with the set of natural numbers as indicator set, we can denote the following event:

$$(X \geqslant 0) = \{\omega \in \Omega \mid X(\omega) \geqslant 0\} = \bigcup_{n \geqslant 0} (X = n) \tag{2.9}$$

The reason for using notation like that in Eqn. (2.9) is merely for presentation purposes. It is used to avoid bloated set notation in formulas. For the same reason, it is also usual to use predicate logic expressions, also over families of random variables. The target is always to achieve formulas that are easier to read. It is common not to introduce such notation formally. Such notation is rather introduced on the fly and should be self-explanatory. For example, given a countable set of random variables $(X_n)_{n \geqslant 0}$ we can use $(\exists n \geqslant 0. X_n = i)$ to denote the following event:

$$(\exists n \geqslant 0. X_n = i) = \bigcup_{n \geqslant 0} (X_n = i) \tag{2.10}$$

Once more see that $(\exists n \geqslant 0. X_n = i)$ in Eqn. (2.10) actually stands for $\{\omega \in \Omega \mid \exists n \geqslant 0 \,.\, X_n(\omega) = i\,\}$. The notations shown in Eqns. (2.9) and (2.10) are concrete examples of notations used in this book, e.g., in the definition of hitting times η in Sect. 2.2.

The indicator set I can be finite, countable or non-denumerable. This is worth mentioning, because the case that the indicator set is $\mathbb{R}$ is so important and convenient that in many textbooks the notion of random variable is introduced immediately and only with $\mathbb{R}$ as the indicator set. In this text we will be interested mostly in the case of a countably infinite indicator set, because we are interested in the terms of a language as an indicator set. In case the indicator set is countable, the random variable is called a discrete random variable.

Next, we recap the notion of expected value of a discrete random variable. Furthermore, we will also need to define the notion of conditional expected value. We need these concepts for the definition of mean reduction lengths of a program; compare with Def. 3.7. For our purposes, it is sufficient to assume that the indicator set of the random variable is the set $\mathbb{N}_0 \cup \{\infty\}$ of natural numbers plus infinity, reflecting the possible reduction lengths before hitting a term. Remember that a random variable is used to assign values to outcomes: in particular, we can use them to weight outcomes of a probability experiment and it therefore makes sense to speak of the values assigned to outcomes as outcome values. Now, informally, the expected value of a random variable is the mean of outcome values of a probability experiment after conducting the experiment a large number of times or, let's say it better, a sufficiently large number of times. In Def. 2.11 we assume, as usual, that $\infty \cdot 0 = 0$, $\infty \cdot n = \infty$ and $n + \infty = \infty$ to write the definitions as sums, instead of writing them as explicit case distinctions.

Definition 2.11 (Expectation, Conditional Expected Value) *Given a random variable* $X : \Omega \rightarrow \mathbb{N}_0 \cup \{\infty\}$ *based on a probability space* $(\Omega, \Sigma, \mathsf{P})$

and an event $A \subseteq \Omega$, we define the expected value $\mathsf{E}(X)$ *of* X*, or expectation of* X *for short, as well as the expected value* $\mathsf{E}_A(X)$ *of* X *conditional on* A *as follows:*

$$\mathsf{E}(X) = \sum_{i=0}^{\infty} i \cdot \mathsf{P}(X = i) + \infty \cdot \mathsf{P}(X = \infty) \tag{2.11}$$

$$\mathsf{E}_A(X) = \sum_{i=0}^{\infty} i \cdot \mathsf{P}_A(X = i) + \infty \cdot \mathsf{P}_A(X = \infty) \tag{2.12}$$

As usual, we also use $\mathsf{E}(X|A)$ to denote $\mathsf{E}_A(X)$. Note that it is very common to use the symbol μ to denote the expected value $\mathsf{E}(X)$. In this book, we use the symbol μ to denote the recursion operators of typed lambda calculi. Therefore, we will avoid using μ to denote expected values in this text.

2.2 Markov Chains

Markov chains are used to model probabilistic, state-changing systems for which the probability of each move only depends on the state of a finitely and absolutely bounded number of prior states. A widespread special case is given by those Markov chains for which the probability of each next move depends solely on the current state. This case is so common that Markov chains are often introduced in the form of this special case. Markov chains in which, furthermore, the next move is independent of the number of steps already taken are called time-homogenous Markov chains. Markov chains that model time as discrete are called discrete-time Markov chains (DTMC). The case that time is modeled as discrete is very common. Therefore, discrete-time Markov chains (DTMC) are usually called Markov chains for short. And, similarly, it is also usual to call time-homogenous Markov chains just Markov chains for short. Also, in this book, we will only deal with time-homogenous, discrete-time Markov chains. Therefore, we feel free to call them Markov chains for short, unless we feel it is important, e.g., in definitions or proofs, to stress the fact that we deal with a time-homogenous, discrete-time Markov chain.

A Markov chain models a state-changing system as a countable infinite family of random variables. The indicator sets of all of these random variables are equal and consist of the states of the system model. Now, the n-th random variable models the probabilities that the system is in each of its states after n steps. Each outcome of the underlying probability space models a state-changing process instance. We are tempted to talk about the outcomes as the processes of the state-changing system. However, it is usual to call the whole system a process, so that it is more accurate to talk about the outcomes as process instances. Similarly, we are tempted to talk about the outcomes as random walks. Again, it is usual to call a whole system a random walk, which is moreover reserved for Markov chains that obey further special conditions. Furthermore, the Markov chain and the system it models are conceptually so close that it is often usual to talk about the modeled system as the Markov chain by which it is modeled. For texts on Markov chains, see, e.g., [211, 181, 177, 179].

In this book we will exploit Markov chains to systematically give an operational semantics to the probabilistic lambda calculus. We do so by identifying the term reduction system of the probabilistic lambda calculus with a concrete Markov chain, in which the terms form the Markov chain states, Markov chain moves correspond to reduction steps and the outcomes of the underlying probability space correspond to concrete program runs.

Definition 2.12 (Time-Homogenous Discrete-Time Markov Chain) *An infinite series, i.e., $\mathbb{N}_0$-indexed family of random variables $(X_n)_{n\geqslant 0}$ with $X_n : \Omega \to S$, is called a time-homogenous discrete-time Markov chain with state space S, or Markov chain for short,* **iff** *the followoing two conditions hold:*

1. *(Markov Property) For all $n \geqslant 0$ and states $s_0, \ldots, s_{n+1}$ it holds that*

$$\mathsf{P}(X_{n+1} = s_{n+1} \mid X_0 = s_0, \ldots, X_n = s_n) = \mathsf{P}(X_{n+1} = s_{n+1} \mid X_n = s_n) \tag{2.13}$$

2. *(Time Homogenity) For all $n, m \geqslant 0$ and states $s, s' \in S$ it holds that*

$$\mathsf{P}(X_{n+1} = s' \mid X_n = s) = \mathsf{P}(X_{m+1} = s' \mid X_m = s) \tag{2.14}$$

The property that the probability of a Markov chain move only depends on a maximum number of prior states is often called the Markov property. Due to the Markov property, a time-homogenous Markov chain can be uniquely characterized by its so-called transition matrix, which contains the probability of a possible move for any two states. Assume that the source states are organized in the rows of the two-dimensional transition matrix, whereas the target states are organized in the columns of the matrix. Now, the transition matrix of a Markov chain is a probability matrix, which means that each row of the transition matrix is a probability distribution; compare with Def. 2.4.

Definition 2.13 (Probability Matrix) *A matrix $m : I \times I \to [0, 1]$ is called a probability matrix* **iff** *each row of the matrix is a probability distribution, i.e., for all $i \in I$ it holds that*

$$\sum_{j\in I} m_{ij} = 1 \tag{2.15}$$

Definition 2.14 (Transition Matrix of a Markov Chain) *Given a time-homogenous Markov chain $X = (X_n : \Omega \to S)_{n\geqslant 0}$. The probability matrix of the Markov chain X, also called the transition matrix of the Markov chain X, is a probability matrix $p : S \times S \to [0, 1]$ on the state set S which is defined for all $i, j \in S$ as follows:*

$$p_{ij} \equiv_{DEF} \mathsf{P}(X_1 = j \mid X_0 = i) \tag{2.16}$$

With $X_0 = i$ we have chosen the start time $n = 0$ as exit time in Def. 2.14, however, this is an arbitrary choice. For a time-homogenous Markov chain, the probability of evolving from a given state i into another given state j is the same independent of the number of steps that have already been taken. This is exactly because of time-homogeneity as expressed by property (2) of Def. 2.12. Therefore, the exit time n in the definition of the transition matrix could be arbitrarily chosen. Time homogeneity can be restated in terms of the transition matrix by the following Corollary 2.15.

Corollary 2.15 (Transition Matrix of a Markov Chain) *Given a time-homogenous Markov chain $X = (X_n : \Omega \to S)_{n \geqslant 0}$ and its transition matrix $p : S \times S \to [0, 1]$. Then for all $n \geqslant 0$ and all $i, j \in S$ it holds that*

$$\mathsf{P}(X_{n+1} = j \mid X_n = i) = p_{ij} \tag{2.17}$$

Proof. This follows immediately from the time-homogenity of the Markov chain X, i.e., property (2) of Def. 2.12. □

Next, we introduce the notation $\mathsf{P}_s(A)$ for the probability of an event conditional on the fact that the Markov chain starts in state s. The introduction of this notation is really only a notational issue. It is nothing but a short-cut notation for the notation $\mathsf{P}_{X_0=s}(A)$. The notation $\mathsf{P}_s(A)$ is introduced because it is a very usual case to consider a probability conditional on a given start state.

Definition 2.16 (Probability Conditional on Start State) *Given a time-homogenous Markov chain $X = (X_n : \Omega \to S)_{n \geqslant 0}$, an event $A \subseteq \Omega$ and a state $s \in S$, we define the probability $\mathsf{P}_s(A)$ of event A conditional on the event that process instances start in s, or conditional on s for short, as follows:*

$$\mathsf{P}_s(A) = \mathsf{P}(A \mid X_0 = s) \tag{2.18}$$

The initial distribution ι is a function that assign the probability $\mathsf{P}(X_0 = i)$ to each state i. We denote the value of ι for state i as ι_i and call it the initial probability of i. Note that ι_i is the probability that a Markov chain is in state i before it undertakes its first step. We can also say that ι_i is the probability that the Markov chain starts in i or, equally well, the probability that i is the start state of the Markov chain. The initial distribution ι actually forms a probability distribution; see Def. 2.4.

Definition 2.17 (Initial Distribution of a Markov chain) *Given a time-homogenous Markov chain $X = (X_n : \Omega \to S)_{n \geqslant 0}$. The initial distribution of the Markov chain X is a probability distribution $\iota : S \to [0, 1]$ on the state set S which is defined for all $i \in S$ as follows:*

$$\iota_i \equiv_{DEF} \mathsf{P}(X_0 = i) \tag{2.19}$$

Sometimes, we denote an initial distribution of a Markov chain ι also as $\vec{\iota}$, in particular, if the initial distibution is used as a row vector in matrix operations. Next, Theorem 2.18 summarizes some crucial properties of Markov chains that all follow from the definition of Markov chains in Def. 2.12. The properties tell us how to compute concrete probabilities on the basis of a probability matrix.

Theorem 2.18 (Properties of Markov Chains) *Given a time-homogenous Markov chain* $X = (X_n : \Omega \to S)_{n \geqslant 0}$ *with state space* S *and transition matrix* $\boldsymbol{P} = p : S \times S \to [0, 1]$, *then for all* $n, m \geqslant 0$ *and* $i, j, i_0, \ldots, i_n \in S$ *it holds that*

1. $\mathsf{P}(X_0 = i_0 \wedge \ldots \wedge X_n = i_n) = \iota_{i_0} \cdot p_{i_0 i_1} \cdot p_{i_1 i_2} \cdot \cdots \cdot p_{i_{n-1} i_n}$
2. $\mathsf{P}(X_n = j) = (\vec{\iota}\boldsymbol{P}^n)_j$
3. $\mathsf{P}(X_n = j \mid X_0 = i) = \mathsf{P}(X_{m+n} = j \mid X_m = i) = (\boldsymbol{P}^n)_{ij}$

2.2.1 Existence of Markov Chains

In Def. 2.14 we have defined the transition matrix for a given Markov chain. Now, we take the opposite direction. Given a probability matrix m, we can be sure that there exists a Markov chain that has m as transition matrix. This greatly eases things in modeling a probabilistic system. It means that there is no need to explicitly construct the outcome space of the underlying probability space, as long as we know the transition probabilities that characterize the probabilistic system under consideration. We make use of this when we define the operational semantics of the probabilistic lambda calculus in Sect. 3.3. We will not fix a concrete set of outcomes, i.e., there is no need to syntactically construct a set of program runs. In a sense, we stay axiomatic. Later, in Chap. 4 we will construct reduction graphs and reduction trees syntactically, because we need a means to exploit graph theory, see Sect. 2.3, to analyze the termination behavior of the probabilistic lambda calculus and to develop crucial propositions with respect to this. Still, reduction graphs and reduction trees are not outcomes themselves but viewports onto the outcomes of the Markov chain semantics.

Theorem 2.19 (Existence of Markov Chains) *Given a probability matrix* $m : S \times S \longrightarrow [0, 1]$ *for an at most countable state set* S *and a probability distribution* $d : S \longrightarrow [0, 1]$. *Then, there exists a Markov chain* $X = (X_n : \Omega_S \to S)_{n \geqslant 0}$ *based on a probability space* $(\Omega_S, \Sigma_S, \mathsf{P})$ *with transition matrix* m *and initial distribution* d, *i.e., for all* $n \in \mathbb{N}_0$ *and* $i, j \in S$ *we have that*

1. $\mathsf{P}(X_{n+1} = j \mid X_n = i) = m_{ij}$
2. $\mathsf{P}(X_0 = i) = d_i$

Proof. No proof provided. See, e.g., §1.3, Theorem 13 and Corollary 14 in [245] or §2.1.1.and Theorem 6.3.2 in [254]. □

It is the theorem of Ionescu-Tulcea that proves the existence of Markov chains even for state spaces that are non-denumerable, i.e., in case $S = \mathbb{R}^n$. We work with countable state sets only in the sequel, so that Theorem 2.19 is sufficient for our purposes.

Corollary 2.20 (Existence of Markov Chains (ii)) *Given a probability matrix $m : S \times S \longrightarrow [0,1]$ for an at most countable state set S. Then, there exists a Markov chain $(X_n)_{n \geqslant 0}$ with transition matrix m and initial distribution ι so that $\iota_s > 0$ for all $s \in S$.*

Proof: We can choose an arbitrary probability distribution $d : S \longrightarrow [0,1]$ with $d_s > 0$ for all $s \in S$ for the given S and apply Theorem 2.19. Corollary 2.20 follows immediately with $\iota = d$. Due to the fact that S is countable, we can assume that $S = \{s_0, s_1, s_2, \ldots\}$. We can, for example, construct the needed probability measure as $s_i \mapsto (0.5)^i$. □

The choice of initial distribution d in Corollary 2.20 is not relevant for the development of the subsequent theory. Ultimately we are not interested in the absolute probabilities of the states of the developed term reduction system, but in the probabilities of states relatively to given start states, which amounts to probabilities dependent on given start states. Therefore the initial distribution always eventually factorizes out. It is only important that the initial distribution is non-zero for all states that we are interested in as start states, so that the probability conditional on these states as start states is well defined. Therefore, we have introduced the side condition $\iota_s > 0$ for all states s in Corollary 2.20.

Definition 2.21 (Markov Chain for Probability Matrix) *Given a probability matrix m and an arbitrary but fixed Markov chain $(X_n)_{n \geqslant 0}$ that has been constructed according to Theorem 2.19 and Corollary 2.20, we use the following notation for the constructed Markov Chain:*

$$\mathfrak{M}(m) =_{DEF} (X_n)_{n \geqslant 0} \tag{2.20}$$

We need $\mathfrak{M}(m)$ as introduced by Def. 2.21 as a succinct notation to define the operational semantics of the probabilistic lambda calculus later.

2.2.2 Hitting Probabilities

Often, we are particularly interested in the so-called hitting probability, which is the probability that a certain state is eventually reached by the process instances of a Markov chain. In our semantics of the probabilistic lambda calculus, we are interested in the probability that a given term ever reduces to another given term, which forms the notion of reduction probability, which is the central notion of our Markov chain semantics. It is a usual option to define the hitting probability via the notion of first hitting time. It is also possible to define this so-called hitting probability directly, without detour, and we will see how in due course in Corollary 2.25. Given a Markov chain X and a set of target states T, the first hitting time is defined as a random variable $H_X(T)$ that assigns, to each process instance ω, the number of steps after which ω hits a state from T for the first time.

Definition 2.22 (First Hitting Time) *Given a Markov chain $X = (X_n : \Omega \longrightarrow S)_{n \geqslant 0}$ and a subset of target states $T \subseteq S$, we define the random variable $H_X(T)$ of first hitting times by*

$$H_X(T) : \Omega \longrightarrow \mathbb{N}_0 \cup \{\infty\} \tag{2.21}$$

$$H_X(T)(\omega) =_{DEF} \begin{cases} \sqcap\{i \in \mathbb{N}_0 \mid X_i(\omega) \in T\} & , \exists i \in \mathbb{N}_0 . X_i(\omega) \in T \\ \infty & , else \end{cases} \tag{2.22}$$

The greatest lower bound notation $\sqcap$ is used here to select the minimum, i.e., smallest number of $\{i \in \mathbb{N}_0 \,|\, X_i(\omega) \in T\}$. First hitting times are also often called first passage times. Next, based on the notion of first hitting times $H_X(T)$ we define the notion of hitting probabilities.

Definition 2.23 (Hitting Probability) *Given a Markov chain $X = (X_n : \Omega \longrightarrow S)_{n \geqslant 0}$, a subset of target states $T \subseteq S$ and a start state s, we define the probability $\eta_X\langle s, T\rangle$ of ever hitting one of the target states from start state s as follows:*

$$\eta_X\langle s, T\rangle =_{DEF} \mathsf{P}_s(H_X(T) < \infty) \tag{2.23}$$

Note that a probability $\eta_X\langle s, T\rangle$ is a conditional probability; see Defs. 2.16 and 2.3. It is the probability of ever hitting one of the target states conditional on the event that the Markov chain starts in state s initially. For instructive purposes, we also give the verbose definition of the hitting probability:

$$\eta_X\langle s, T\rangle = \frac{\mathsf{P}(H_X(T) < \infty \ \wedge \ X_0 = s)}{\mathsf{P}(X_0 = s)} \tag{2.24}$$

Given a single target state t, we also write $\eta_X\langle s, t\rangle$ to denote $\eta_X\langle s, \{t\}\rangle$. In proofs we are often interested in a hitting probability up to a given maximum number of steps. We call such a kind of probability a bounded hitting probability.

Definition 2.24 (Bounded Hitting Probability) *Given a Markov chain $X = (X_n : \Omega \longrightarrow S)_{n \geqslant 0}$, a subset of target states $T \subseteq S$, a start state s and an upper bound $n \geqslant 0$, we define the probability $\eta^n_X\langle s, T\rangle$ of hitting one of the target states from start state s in at most n steps as follows:*

$$\eta^n_X\langle s, T\rangle =_{DEF} \mathsf{P}_s(H_X(T) \leqslant n) \tag{2.25}$$

Again, given a single target state t, we also write $\eta^n_X\langle s,t\rangle$ to denote $\eta^n_X\langle s,\{t\}\rangle$. Now, we introduce alternative, more direct, definitions of hitting probabilities and bounded hitting probabilities without the detour via first hitting times, in Corollary 2.25. The notation used in the Corollary to specify the respective events in Eqns. (2.26) and (2.27) has already been discussed in Sect. 2.1.3; see Eqn. (2.10) and the surrounding discussion.

Corollary 2.25 (Alternative Hitting Probability Definitions) *Given a Markov chain $X = (X_n : \Omega \longrightarrow S)_{n\geqslant 0}$, a subset of target states $T \subseteq S$ and a start state s the following characterizations hold true:*

$$\eta_X\langle s,T\rangle = \mathsf{P}_s(\exists\, i<\infty \,.\, X_i\in T) \tag{2.26}$$

$$\eta^n_X\langle s,T\rangle = \mathsf{P}_s(\exists\, i\leqslant n \,.\, X_i\in T) \tag{2.27}$$

Proof. The corollary follows from the definition of first hitting times in Def. 2.22, hitting probabilities in Def. 2.23 and bounded hitting probabilities in Def. 2.24. □

The alternative definitions of hitting probabilities in Corollary 2.25 are more straightforward than the ones via hitting probabilities given in Defs. 2.23 and 2.24 and, actually, they are even more intuitive. Nevertheless, the definition of hitting probabilities via hitting times can often be found in textbooks and turns out to be useful in inductive proofs, because first hitting times enjoy the property that they are disjoint for different numbers of steps, i.e.,

$$i \neq j \implies (H_X(T) = i) \cap (H_X(T) = j) = \emptyset \tag{2.28}$$

Furthermore, the random variable $H_X(T)$ is crucial for the definition of mean hitting times in Def. 2.32. So, we gain a certain correspondence in the definition of hitting probabilities and mean hitting times when the definition of hitting probabilities is based on the random variable of first hitting times. Several further characterizations of hitting probabilities and bounded hitting probabilities exist that can also be very useful in proofs; see Corollaries 2.26 and 2.27.

Corollary 2.26 (Characterizations of Hitting Probabilities) *Given a Markov chain $X = (X_n : \Omega \longrightarrow S)_{n\geqslant 0}$, a subset of target states $T \subseteq S$, a start state s and a number $n \in \mathbb{N}_0$ the following characterizations hold true:*

$$\eta_X\langle s,T\rangle = \sum_{n=0}^{\infty} \mathsf{P}_s(H_X(T) = n) \tag{2.29}$$

$$\eta^n_X\langle s,T\rangle = \sum_{i=0}^{n} \mathsf{P}_s(H_X(T) = n) \tag{2.30}$$

Proof. This follows from the additivity of the probability function and the fact that $H_X(T)=n$ and $H_X(T)=m$ are disjoint for all $m \neq n$. □

Corollary 2.27 (Limit of Bounded Hitting Probabilities) *Given a Markov chain $X = (X_n : \Omega \longrightarrow S)_{n \geqslant 0}$, a subset of target states $T \subseteq S$, a start state s and numbers $m, n \in \mathbb{N}_0$ the following holds true:*

$$\eta_X\langle s,T\rangle = \lim_{n\to\infty} \eta_X^n\langle s,T\rangle \tag{2.31}$$

Proof. Immediate from Corollary 2.26. □

Bounded hitting probabilities increase monotonically with the bound of steps as expressed by the following Corollary 2.28.

Corollary 2.28 (Monotonicity of Bounded Hitting Probabilities) *Given a chain Markov $X = (X_n : \Omega \longrightarrow S)_{n \geqslant 0}$, a subset of target states $T \subseteq S$, a start state s and numbers $m, n \in \mathbb{N}_0$ the following holds true:*

$$m \geqslant n \implies \eta_X^m\langle s,T\rangle \geqslant \eta_X^n\langle s,T\rangle \tag{2.32}$$

Proof. Immediate from Corollary 2.26. □

The vector of hitting probabilities of a Markov chain can be characterized as the least solution of a linear equation system – see Theorem 2.29. The characterization as a linear equation system is important for us, because, based on it, we can exploit the results and algorithms of linear algebra to determine and argue about reduction probabilities of the probabilistic lambda calculus.

Theorem 2.29 (Linear Equation System of Hitting Probabilities) *Given a Markov chain $X = (X_n : \Omega \longrightarrow S)_{n \geqslant 0}$ with transition matrix $p : S \times S \to [0,1]$, a start state s and a set of target states $T \subseteq S$. The vector of hitting probabilities $(\eta_X\langle i,T\rangle)_{i\in S}$ is the least solution of the following equation system:*

$$\begin{aligned} &\forall s \in T\,. && \eta_X\langle s,T\rangle = 1 \\ &\forall s \notin T\,.\big(\sum_{t\in S} p_{st}\cdot \eta_X\langle t,T\rangle\big) && - \eta_X\langle s,T\rangle = 0 \end{aligned} \tag{2.33}$$

Proof. See, e.g., Theorem 1.3.2 in [211], or Theorem 1.24 in [177]. □

All the terms of the form $\eta_X\langle s,T\rangle$ as well as all terms of the form $\eta_X\langle t,T\rangle$ for each state $s \in S$ resp. $t \in S$ stand for individual variables in the equation system of Theorem 2.29. The equation system consists of an equation for each of the states in S. Therefore, in general, the equation system is infinite. Later, it will be a crucial step to reduce the state set of the considered Markov chain to a finite subset, so that linear algebra algorithms become applicable. The linear equation system in Eqn. (2.33) is in matrix form $\mathbf{M} \times \mathbf{x} = \mathbf{r}$ with a column vector $\mathbf{r}$ of scalar values. However, the correctness of Theorem 2.29 is easier to see if the equation system is rewritten in a slightly different form:

$$\forall s \in T . \eta_X\langle s,T\rangle = 1 \tag{2.34}$$

$$\forall s \notin T . \eta_X\langle s,T\rangle = \sum_{t\in S} p_{st} \cdot \eta_X\langle t,T\rangle \tag{2.35}$$

A set of Markov chain states is called a closed class if it is not possible to escape from it, once a state of the set has been reached. A hitting probability $\eta_X\langle s,T\rangle$ is called an absorption probability if the set T is closed. The states of a closed class are also called absorbing states. Given an absorbing state s and a process instance ω we say that ω is absorbed behind s. In our lambda calculus semantics we deal primarily with hitting probabilities that are absorption probabilities. We usually consider cases in which the set of target states consists of a single state and the Markov chain henceforth remains cycling in this state.

Definition 2.30 (Closed Class) *Given a Markov chain* $X = (X_n : \Omega \longrightarrow S)_{n\geqslant 0}$ *and a subset of target states* $T \subseteq S$, *the set* T *is called a closed class of states, or closed for short,* **iff** *all states* $s' \in S$ *that are reachable from a source state* $s \in T$ *also belong to* T, *i.e.,*

$$\forall s \in T \,.\, \forall s' \in S \,.\, (\ \eta_X\langle s,s'\rangle > 0 \implies s' \in T\ ) \tag{2.36}$$

The fact that hitting probabilities are characterized by the equations in Eqns. (2.34) and (2.35) is important because it allows us to determine concrete hitting probability values. It is also important because the equations are needed to prove propositions about Markov chains and the system they model. For us, a similar result about bounded hitting probabilities is important, which is expressed in the decomposition Lemma 2.31. Lemma 2.31 explains how a bounded hitting probability can be calculated from the bounded hitting probabilities of a fewer number of steps.

Lemma 2.31 (Decomposition of Bounded Hitting Probabilities) *Given a Markov chain* $X = (X_n : \Omega \longrightarrow S)_{n\geqslant 0}$ *with transition matrix* $p : S \times S \to [0,1]$ *and a bound* $n \in \mathbb{N}_0$ *the following characterizations of bounded hitting probabilities hold true:*

$$\forall s \in T . \eta_X^n\langle s,T\rangle = 1 \tag{2.37}$$

$$\forall s \notin T . \eta_X^0\langle s,T\rangle = 0 \tag{2.38}$$

$$\forall s \notin T . \eta_X^{n+1}\langle s,T\rangle = \sum_{t\in S} p_{st} \cdot \eta_X^n\langle t,T\rangle \tag{2.39}$$

Proof. Let us start with Eqn. (2.37), which follows from the definition of bounded hitting probabilities in Def. 2.24. Due to Eqn. (2.27) we have that $\eta_X^n\langle s,T\rangle$ equals

$$\mathsf{P}(\exists i \leqslant n. X_i \in T \wedge X_0 = s) \,/\, \mathsf{P}(X_0 = s) \tag{2.40}$$

Due to the premise of Eqn. (2.37), i.e., $s \in T$, we have that $(X_0 = s)$ is a subset of $(\exists i \leqslant n. X_i \in T)$. Therefore, we know that Eqn. (2.40) equals $\mathsf{P}(X_0 = s)/\mathsf{P}(X_0 = s)$ which equals 1.

Let us turn to Eqn. (2.38). Again, due to Eqn. (2.27) we have that $\eta_X^0\langle s, T\rangle$ equals

$$\mathsf{P}(\exists i \leqslant 0. X_i \in T \wedge X_0 = s) \,/\, \mathsf{P}(X_0 = s) \tag{2.41}$$

We know that $(\exists i \leqslant 0. X_i \in T)$ equals $(X_0 \in T)$. Now, given that $s \notin T$ we know that $(X_0 \in T) \cap (X_0 = s) = \emptyset$. Due to $\mathsf{P}(\emptyset) = 0$ we have that Eqn. (2.41) equals 0.

Now, let us show Eqn. (2.39). Due to Eqn. (2.27) we have that $\eta_X^{n+1}\langle s, T\rangle$ equals

$$\mathsf{P}(\exists\, i \leqslant n{+}1 \,.\, X_i \in T \wedge X_0 = s) \,/\, \mathsf{P}(X_0 = s) \tag{2.42}$$

Due to the premise that $s \notin T$ and similar arguments as in case of Eqn. (2.38) we have that Eqn. (2.42) equals

$$\mathsf{P}(\exists\, 1 \leqslant i \leqslant n{+}1 \,.\, X_i \in T \wedge X_0 = s) \,/\, \mathsf{P}(X_0 = s) \tag{2.43}$$

Now, due to the countable additivity of the probability function, see Def. 2.2, and the fact that $(X_1 {=} t)$ is disjoint from $(X_1 {=} t')$ for all $t {\neq} t'$ we have that Eqn. (2.43) equals

$$\sum_{t \in S} \mathsf{P}(\exists\, 1 \leqslant i \leqslant n{+}1 \,.\, X_i \in T \wedge X_0 = s \wedge X_1 = t) \,/\, \mathsf{P}(X_0 = s) \tag{2.44}$$

Now, Eqn. (2.44) can always be transformed into the following:

$$\sum_{t \in S} \frac{\mathsf{P}(X_0 = s \wedge X_1 = t)}{\mathsf{P}(X_0 = s)} \cdot \frac{\mathsf{P}(\exists\, 1 \leqslant i \leqslant n{+}1 \,.\, X_i \in T \wedge X_0 = s \wedge X_1 = t)}{\mathsf{P}(X_0 = s \wedge X_1 = t)} \tag{2.45}$$

Just for convenience, let us rewrite Eqn. (2.45) as follows:

$$\sum_{t \in S} \mathsf{P}(X_1 = t \mid X_0 = s) \cdot \mathsf{P}(\exists\, 1 \leqslant i \leqslant n{+}1 \,.\, X_i \in T \mid X_0 = s, X_1 = t) \tag{2.46}$$

Now, due the definition of transition matrices of Markov chains in Def. 2.14, we have that Eqn. (2.46) equals the following:

$$\sum_{t \in S} p_{st} \cdot \mathsf{P}(\exists\, 1 \leqslant i \leqslant n{+}1 \,.\, X_i \in T \mid X_0 = s, X_1 = t) \tag{2.47}$$

Next, due to the Markov property, i.e., Eqn. (2.13), it can be shown that Eqn. (2.47) equals the following:

$$\sum_{t \in S} p_{st} \cdot \mathsf{P}(\exists\, 1 \leqslant i \leqslant n{+}1 \,.\, X_i \in T \mid X_1 = t) \tag{2.48}$$

Due to the time homogeneity of the involved Markov chain, i.e., Eqn. (2.14), it can be shown that Eqn. (2.48) equals the following:

$$\sum_{t \in S} p_{st} \cdot \mathsf{P}(\exists\, 0 \leqslant i \leqslant n \,.\, X_i \in T \mid X_0 = t) \tag{2.49}$$

Finally, due to the definition of bounded hitting probabilities in Def. 2.24 we have that Eqn. (2.49) equals:

$$\sum_{t \in S} p_{st} \cdot \eta_X^n \langle t, T \rangle \tag{2.50}$$

□

Next, we define the mean hitting time $|\eta_X\langle s, T\rangle|$ of reaching one of the target states in T from a start state s. It is defined as the expected average number of steps needed to reach one of the target states from the start state.

Definition 2.32 (Mean Hitting Time) *Given a Markov chain $X = (X_n : \Omega \longrightarrow S)_{n \geqslant 0}$, a subset of target states $T \subseteq S$ and a start state $s \in S$, we define the mean hitting time $|\eta_X\langle s, T\rangle|$ needed to reach a state in T starting from s as follows:*

$$|\eta_X\langle s, T\rangle| =_{DEF} \mathsf{E}_s(H_X(T)) \tag{2.51}$$

Given a single target state t, we also write $|\eta_X\langle s, t\rangle|$ to denote $|\eta_X\langle s, \{t\}\rangle|$. The notion of mean hitting time is based on the random variable of first hitting times defined in Def. 2.22 and the notion of conditional expectation in Def. 2.11. Again, mean hitting times can be determined as a solution of a linear equation system; see Theorem 2.33. Again, each term of the form $|\eta_X\langle s, T\rangle|$ in Eqn. (2.52) stands for an individual variable.

Theorem 2.33 (Linear Equation System of Mean Hitting Times) *Given a Markov chain $X = (X_n : \Omega \longrightarrow S)_{n \geqslant 0}$ with transition matrix $p : S \times S \to [0, 1]$, a start state s and a set of target states $T \subseteq S$. The vector of mean hitting times $(|\eta_X\langle i, T\rangle|)_{i \in S}$ is the least solution of the following equation system:*

$$\begin{aligned} &\forall i \in T \,. && |\eta_X\langle i, T\rangle| = 0 \\ &\forall i \notin T \,. \Big(\sum_{j \notin T} p_{ij} \cdot |\eta_X\langle j, T\rangle|\Big) && -|\eta_X\langle i, T\rangle| = -1 \end{aligned} \tag{2.52}$$

Proof. See, e.g., Theorem 1.3.5 in [211]. □

2.3 Graph Theory

We need graph theory for the analysis of the operational semantics of the probabilistic lambda calculus in Sect. 4. We are particularly interested in the termination behavior of probabilistic programs. We will establish the notion of unbounded termination. We say that a program terminates unbounded if it has terminating program runs of arbitrary length. We will see that an unbounded program necessarily also has a non-terminating program run. We will identify a class of probabilistic programs for which the termination degree is computable, the so-called path stoppable programs. All of these results rely on a graph-theoretic treatment of the reduction system of the probabilistic lambda calculus. Basically, we will view the transition matrix of our Markov chain semantics as a graph, the so-called reduction graph, in order to achieve the notion of path stoppability. Only after the establishment of reduction graphs we step further to define reduction trees, which become necessary in the analysis of bounded vs. unbounded terminating programs. In order to prove the desired properties of program behavior we exploit graph-theoretic results, i.e., König's Lemma and Beth's Theorem.

In order to establish the notion of path stoppability and to prove the main result for it, i.e., the computability of termination behavior for a certain class of programs, we need to prove further lemmas for graphs. We need the result that the nodes that are covered by the walks of a graph are already completely covered by the paths of the graph. Furthermore, we need to state under what circumstances the cover of a graph is computable. These two important lemmas are not proven here in this chapter but are deferred until Chap. 4 where they occur as Lemma 4.25 and Theorem 4.26 in the context where they naturally emerge. This shows, once more, the appropriateness of a graph-theoretic viewpoint of the operational semantics.

2.3.1 Digraphs

In this book we work with digraphs, i.e., directed graphs. Often, we call digraphs also graphs for short, in particular if there is no risk of confusion or we do not want to stress explicitly that we deal with directed graphs. Our notion of digraph is straightforward. There is at most one edge between two vertices, i.e., our digraphs are not multigraphs. This notion of graph is particularly appropriate for our purpose of modeling the reduction graph and, with its edges being modeled as a relation, particularly easy to handle in proofs. From time to time it is instructive to consider digraphs as graphs in general; this is why we also introduce the definitions of graphs in general later in this section. See [16] for a thorough treatment of digraphs.

Definition 2.34 (Digraph) *A digraph (directed graph) G is a tuple $\langle V, E\rangle$ consisting of vertices V, also called nodes, and edges $E \subseteq V \times V$.*

A digraph $G = \langle V_G, E_G\rangle$ is considered to be directed, because its edges are ordered pairs. The direction is somehow considered from left to right as shown in the definition of walks and paths later. Furthermore, given an edge $e = \langle v, v'\rangle$ between nodes v and v', the edge e is said to be an outgoing edge of v and, at the same time, it is said to be an ingoing edge of v'. The number of outgoing edges of a node is called the degree of that node. If the number of outgoing edges of a digraph is limited by a number k for all nodes of a digraph G, the digraph G is said to be k-ary. The degree of a node can be either finite or infinite; we then also say that the node is of finite degree resp. infinite degree. Next, we introduce the notion of walks in a digraph. A walk in a digraph is a sequence of nodes that are, sequentially, connected by edges as defined in Def. 2.35.

Definition 2.35 (Walks of a Digraph) *Given a digraph* $G = \langle V_G, E_G \rangle$, *the set* nW_G *of its walks of length* $n \in \mathbb{N}_0$, *the set* $\star W_G$ *of its walks of finite length, the set* ωW_G *of its infinite walks and the set* $\oplus W_G$ *of its walks, are defined as sets as follows:*

$$nW_G = \{w \mid w \in V_G^{n+1},\ \forall i < n. \langle w_i, w_{i+1} \rangle \in E_G\} \tag{2.53}$$

$$\star W_G = \{w \mid w \in V_G^{\star},\ \forall i < \#(w) - 1. \langle w_i, w_{i+1} \rangle \in E_G\} \tag{2.54}$$

$$\omega W_G = \{(w_i)_{i \in \omega} \mid w_i \in V_G,\ \forall i \in \mathbb{N}_0. \langle w_i, w_{i+1} \rangle \in E_G\} \tag{2.55}$$

$$\oplus W_G = \star W_G \cup \omega W_G \tag{2.56}$$

If w is a walk of a graph G, we also say that w is a walk in the graph G and we also say G has the walk w. For the definition of V_G^n and $V_G^\star$ see the definition of sequences of length n as well as sequences of arbitrary length in Sect. 2.5.2. As usual, we also use $w_0 w_1 w_2 \cdots w_n$ to denote a walk $(w_i)_{i \in \{0,\ldots,n\}}$. Textbooks [16, 32, 33] on graphs usually introduce walks as sequences $v_0 e_0 v_1 e_1 v_2 e_2 \cdots e_{n-1} v_n$ of interchanging vertices v_i and edges e_i, and so we will do in case of graphs later in Def. 2.40. In case of digraphs, however, we can forget about the edges in the walks. This is possible, because a digraph is not a multigraph, and the edges of its walks are therefore uniquely determined by the vertices of its walks. Note that also textbooks on digraphs such as [16] might use a definition based on sequences involving edges. We prefer a definition based on sequences of vertices only. Then, as opposed to many textbooks, we make explicit the technical details of the sequences in our definition. Last but not least, as an important technical difference, we also deal explicitly with the possibility of infinite walks. All of this is extra accurate but pays back later in proofs, where it shortens notation and eases argumentation. Next, we introduce resp. recap some useful notation for the nodes and sizes of graphs, walks as well as sets of walks.

Definition 2.36 (Nodes and Sizes in Digraphs) *Given a digraph* $G = \langle V_G, E_G \rangle$. *We denote the set of its nodes also by* $\kappa(G)$, *i.e.,* $\kappa(G) = V_G$. *We denote its size by* $|G|$, *i.e.,* $|G| = |V_G|$. *Given a walk* $w \in \oplus W_G$, *we denote the set of its nodes by* $\kappa((w_i)_{i \in I}) = \{w_i \mid i \in I\}$, *its size by* $|w| = |\kappa(w)|$ *and its length by* $l(w) = \#(w) - 1$. *Given a set of walks* $U \subseteq \oplus W_G$, *we denote the set of its nodes by* $\kappa(U) = \cup\{\kappa(u) | \ u \in U\}$ *and its size by* $|U| = |\kappa(U)|$.

A walk consists of at least one node; compare with Eqn. (2.53). Therefore, walks of length zero also consist of one node. We distinguish between the length $l(w)$ of a walk w and its sequence length $\#(w)$. The length of a walk equals its sequence length minus one, which is in accordance with the definition of walks in graphs in Def. 2.40, where the length of a walk is defined as the number of its edges. Note that the size of a walk $w \in \oplus W_G$ is less than or equal to its sequence length, i.e., $|w| \leqslant \#(w) = l(w) + 1$.

Often, we are interested in walks that start with a given node. We introduce some convenient notation for sets of walks that all start with the same node in Def. 2.37. The notation always relies on a given set of walks W. Then we use $W(v)$ to denote the subset of all walks of W that start with v. For example, given a digraph G and a node $v \in \kappa(G)$ the set of all of its walks starting in v is denoted by $\oplus W_G(v)$.

Definition 2.37 (Walks with a Start Node) *Given a digraph G, a set of walks $W \subseteq \oplus W_G$, and a node $v \in \kappa(G)$, we define the set $W(v)$ of walks from W starting in v as follows:*

$$W(v) = \{w \in W \mid w_0 = v\} \tag{2.57}$$

Next, we define the important notion of paths. Informally, a path is a walk that has no cycles, i.e., a walk along which it is not possible to visit a node twice.

Definition 2.38 (Paths) *Given a digraph, a walk is called a path if all of its nodes are pairwise disjoint. Given a digraph G and a set of walks $W \subseteq \oplus W_G$, we define the set $\pi(W)$ of paths of W as follows:*

$$\pi(W) = \{(w_i)_{i \in I} \in W \mid \forall i, j \in I, i \neq j \; . \; w_i \neq w_j\} \tag{2.58}$$

For example, given a digraph G, the set of all of its paths is $\pi(\oplus W_G)$ and the set of all of its paths of length n is $\pi(nW_G)$. Note that the size of a path $p \in \pi(nW_G)$ equals its sequence length, i.e., $|p| = \#(p) = l(p) + 1$.

In acyclic graphs, and in particular in trees, all the walks are paths.

Now, let us turn to König's Lemma. Informally, it states that an infinite graph necessarily has an infinite path, as long as all of its nodes only have finitely many outgoing edges.

Lemma 2.39 (König's Lemma for Digraphs and Start Nodes) *Given a digraph $G = \langle V_G, E_G \rangle$ so that all of its nodes have a finite degree. Given a node $v \in V_G$. If the set of nodes covered by walks from G starting in v is infinite, then there exists an infinite path starting from v, i.e.,*

$$|\kappa(\star W_G(v))| = \infty \implies \exists \omega \; . \; \omega \in \pi(\omega W_G(v)) \tag{2.59}$$

Proof. This Lemma can be proven as a corollary of König's Lemma for graphs, i.e., Theorem 2.42, based on the fact that the sub-digraph consisting of nodes reachable from v, interpreted as a graph, is a connected graph.

2.3.2 Graphs

Like digraphs, graphs consist of nodes that are connected by edges. Unlike digraphs, the edges in a graph are not directed. We recap the definition of ordinary graphs and their walks, trails and paths in Def. 2.40; compare with [32, 70, 33].

Definition 2.40 (Graphs, Walks, Trails, Paths, Degrees) *A **graph** G is a triple $\langle V, E, \phi \rangle$ consisting of a set of vertices V, a set of edges E and a so-called incident function $\phi : E \to \{\{v_1, v_2\} \mid v_1 \in V, v_2 \in V\}$ that assigns an unordered pair of vertices to each edge. A **walk** (infinite walk) of a graph G is a sequence $v_0 e_0 v_1 e_1 \cdots e_{n-1} v_n$ of arbitrary length n resp. an infinite sequence $v_0 e_0 v_1 e_1 \cdots$ of interchanging nodes v_i and edges e_i so that for all $i \geqslant 0$ we have that $\phi(e_i) = \{v_i, v_{i+1}\}$. A **trail** is walk in which no edge occurs more than once. A **path** is a walk in which no node occurs more than once. The **degree** of a node $v \in V$ is defined as the number of its connected edges, i.e., as $|\{e \mid v \in \phi(e)\}|$ in case there are finitely many connected edges, and otherwise, as infinite (∞) in case there are infinitely many connected edges.*

Note that according to Def. 2.40 the length of a path equals the length of its edges. Note that each path of a graph is a trail, but not vice versa. The definition of graphs based on an incident function as defined in Def. 2.40 is standard in the literature, although many other equivalent definitions of course exist. For a detailed treatment of graphs, see [33]. The graphs as defined in Def. 2.40 are undirected multigraphs, i.e., the edges do not have a designated source or target, and there may exist more than one edge between two given vertices.

Given a digraph $D = \langle V, E \rangle$, the digraph can be immediately considered as a graph $\langle V, E, \phi \rangle$ with $\phi(\langle v_i, v_j \rangle) = \{v_i, v_j\}$ for each $\langle v_i, v_j \rangle \in E$. Note that this transformation of a digraph D into a graph loses information, i.e., the information about the direction of edges. Furthermore, the transformation may result in a multigraph. Nevertheless, this re-interpretation of digraphs allows for the exploitation of some results that are available for graphs in general, e.g., König's Lemma in Theorem 2.42.

Definition 2.41 (Connected Graph) *A graph $G = \langle V, E, \phi \rangle$ is connected if each pair $v \neq v' \in V$ of its nodes is connected by a path, i.e., for all pairs $v \neq v' \in V$ there exists some path $p_0 e_0 p_1 e_1 \cdots e_{n-1} p_n$ for some $n \in \mathbb{N}_0$ so that $p_0 = v$ and $p_n = v'$.*

Theorem 2.42 (König's Lemma) *Given a connected graph G so that all vertices of G have finite degree. If G is infinite then there exists an infinite path in G.*

Proof. See [165, 166]. The proof can be conducted by structural induction on the length of walks, by proving the possibility of a next admissible move by an indirect argument as induction step – see also [171]. □

The converse of König's Lemma trivially holds. Each graph that has an infinite path necessarily has infinitely many nodes. This is so, because the nodes in a path are all pairwise disjoint.

2.3.3 Trees

Trees are special graphs. A tree is an acyclic graph in which there exists a unique path between any two of its nodes. In this book we work with directed trees, i.e., trees that are defined on the basis of directed graphs. More concretely, we work with rooted directed trees, which have a designated node as so-called root. Again, we often simply talk about trees instead of rooted directed trees in the sequel, as long as everything is clear from the context and an explicit distinction is not required. First, we introduce the notion of directed acyclic graphs (DAG) in Def, 2.43, then we define rooted directed trees as special DAGs. The definition of leaves and inner nodes of a rooted directed tree in Def. 2.44 is straightforward.

Definition 2.43 (DAGs) *A digraph G is a DAG (directed acyclic graph)* **iff** *all of its walks are paths, i.e., $\star W_G = \pi(\star W_G)$.*

Obviously, a cycle is a walk that eventually returns to its start node. See Def. 2.46 for a definition of cycles in graphs. A similar definition can be given for cycles in digraphs. Therefore, according to Def. 2.43, a DAG contains no cycles; in particular, a DAG also does not contain any self-cycles, i.e., cycles of length one. Note that walks of length zero are considered not to be cycles. Furthermore, walks of length zero are considered to be paths; compare with Def. 2.38. Therefore, walks of length zero do not pose a problem with respect to Def. 2.43 and the notion of DAG.

Definition 2.44 (Rooted Directed Tree) *A rooted directed tree $T = \langle V, E, r\rangle$, rooted tree or tree for short, is a triple that is established by a digraph $\langle V, E\rangle$ and a root node $r \in V$, called its root for short,* **iff** *T is a DAG (directed acyclic graph), and each node different from the root is (end-to-end) connected to the root by exactly one path.*

Definition 2.45 (Leaves and Inner Nodes of a Tree) *Given a rooted tree $T = \langle V, E, r\rangle$ a node $v \in V$ is a leaf* **iff** *it has no outgoing edges, i.e., $\{ v' \mid \langle v, v'\rangle \in E\} = \emptyset$. A node $v \in V'$ is an inner node* **iff** *v is not a leaf.*

Again, for the sake of completeness, we also include here definitions for ordinary trees, i.e., trees that are based on ordinary graphs; again, compare with [32, 33]. We do so because in the sequel we want to exploit results that are available for ordinary trees and immediately transport them to directed trees. We proceed, as usual, by the definition of cycles via the definition of acyclic graphs to the definition of trees and close with the observation that paths between any two nodes of a tree are unique.

Definition 2.46 (Cycle) *Given a graph G, a walk $v_0e_0v_1e_1\ldots v_{n-1}e_{n-1}v_n$ of G is a cycle* **iff** *it is not empty, its first node is equal to its last node and the walk up to the last node is a path, i.e., $n \geqslant 1$, $v_0 = v_n$ and $v_0e_0v_1e_1\ldots e_{n-1}v_{n-1}$ is a path.*

Definition 2.47 (Acyclic Graph) *A graph $G = \langle V, E, \phi\rangle$ is acyclic* **iff** *it contains no cylces.*

Definition 2.48 (Trees) *A tree T is a graph $T = \langle V, E, \phi\rangle$ that is acyclic and connected.*

Lemma 2.49 (Uniqueness of Paths in Tree) *If a graph $T = \langle V, E, \phi\rangle$ is a tree then any two of its nodes $u, w \in V$ are (end-to-end) connected by exactly one path, i.e., there exists exactly one path $v_0e_0v_1e_1\cdots e_{n-1}v_n$ with $v_0 = u$ and $v_n = w$.*

Proof. See Theorem 2.1 in [32]. □

Any rooted directed tree can be re-interpreted as a tree. This is so, because any rooted directed tree is a digraph and therefore can be transformed into a graph as described in Sect. 2.3.2, where it can be shown that the rooted-tree property is preserved by the transformation, i.e., implies the general tree property. This allows for the application of results that are available for trees in general, e.g., Beth's Tree Theorem 2.50, to rooted directed trees. Beth's Theorem can be considered to be an instance of Königs Lemma for the case of trees.

Theorem 2.50 (Beth's Tree Theorem) *Given a tree T. If G is infinite, there exists an infinite path in T.*

Proof. This Theorem is a corollary of König's Lemma, i.e., Theorem 2.42. Apart from that, Beth's Tree Theorem has been found and proven independently of König's Lemma [26, 27]. □

2.4 Inductive Definitions

Inductive definitions are one of the most basic tools in building arbitrary kinds of mathematical structures and arguing about them. In computer science, we come across inductive definitions everywhere, implicitly or explicitly, e.g., always in the definition of programming languages, definition of type systems [42] as well as the investigation of algorithms [52] and formal languages [137]. Usually, the mechanism of inductive definition is just taken for granted. Usually, it is not explained and often it is even not explicitly mentioned. This is so, because the focus regularly needs to be solely on the genuine mathematical or otherwise-structured object of interest. It is different in this book. We make a concrete language of randomized algorithms, i.e., the probabilistic lambda calculus, the subject of our investigation and formal treatise. Therefore, we need to make the theory of inductive definitions explicit, because we need its propositions in conclusions, argumentations and proofs. This is so later when we come up with the denotational semantics for our language. But also earlier, when we treat the operational semantics, it makes sense to make explicit and reflect upon the inductive definitions of the established language. Not everywhere, but in large parts of the book, we have chosen ω-complete partial orders as domains for our inductive definitions. The concrete choice of underlying mathematical structures may always, at least to a certain extent, appear to be arbitrary or a matter of taste. Actually, the choice depends on the needs of the targeted analysis and the requirements of the object of investigations. We aim at motivating our concrete choices of mathematical structures in place here in this chapter but also in the upcoming Chap. 5 where we deal with the denotational semantics. In this chapter we furthermore present the basic motivations of inductive definitions and also delve into technical issues, in particular, concerning ω-cpo*s* and ω-continuous functions.

So what is an inductive definition? Whenever you have defined a set of objects as the least set that fulfills a collection of conditions, you have provided an inductive definition. Actually, the concrete specification of the involved collection of conditions is exactly what inductive definition is about. As a standard example, a formal language can be considered the least set of words that is closed under the production rules of the corresponding grammar. The notion of inductive definition can be brought into shape with the Knaster-Tarski fixed-point theorem. In its earliest version [160] the Knaster-Tarski fixed point theorem is formulated for each power set over a set, ordered by subset inclusion. It is proven that each monotone function over a power set has a least fixed point. Now, the least fixed point of such a monotone function provides an inductive definition. Later [255], the Knaster-Tarski fixed point theorem was formulated more abstractly, in terms of arbitrary complete lattices.

We proceed as follows. We give the necessary definitions for partial orders and lattices needed for the Knaster-Tarski theorem in its generalized form and state this theorem in Sect. 2.4.1. Then, we turn to rule-based inductive definitions in Sect. 2.4.2. Rule-based inductive definitions provide a concrete form, i.e., a concrete notation, to present inductive definitions. Rule-based inductive definitions, or rule-based definitions for short, are particularly intuitive and in widespread use, in particular, in the computer science literature. Nonetheless, it is important to carve out their formal meaning, in order to achieve rigor in the material in which we use them. We will use rule-based inductive definitions in many places, e.g., in the definition of the one-step semantics as well as in the definition of syntactical approximation for lambda terms. Next, we turn to cpo*s* and ω-cpo*s* in Sects. 2.4.3 through 2.4.8. We need to treat ω-cpo*s* in detail, because they are the domains of choice for our denotational semantics. We also review cpo*s* in this book for the sake of completeness and for better comparability with the literature. With this direction and these techniques we have chosen a concrete viewpoint on the field of inductive definitions. For a comprehensive treatment of inductive definitions we also recommend [201, 202].

2.4.1 The Knaster-Tarski Fixed-Point Theorem

A partially ordered set, or partial order, is a set together with a reflexive, anti-symmetric and transitive binary relation. Partial orders are the basis for all further structures that we need in inductive definitions and later in denotational semantics, i.e., lattices, complete lattices, cpos and ω-cpos.

Definition 2.51 (Partial Order) *A partially ordered set $(S, \sqsubseteq)$, also partial order for short, consists of a set S and binary relation $\sqsubseteq$: $S \times S$, which is reflexive, anti-symmetric and transitive, i.e., for all $s, s_1, s_2, s_3 \in S$ it holds that*

1. $s \sqsubseteq s$ *(reflexivity)*
2. $s_1 \neq s_2 \implies \neg(s_1 \sqsubseteq s_2 \wedge s_2 \sqsubseteq s_1)$ *(anti-symmetry)*
3. $s_1 \sqsubseteq s_2 \wedge s_2 \sqsubseteq s_3 \implies s_1 \sqsubseteq s_3$ *(transitivity)*

Given a partial order $D = (D', \sqsubseteq_{D'})$ as defined by Def. 2.51. Then, the set D' is also called the carrier or base set of D. We also say d is an element of D if d is an element of the carrier D', and also write $d \in D$ for $d \in D'$. Often, in terms of used symbols, we want to neglect the distinction between a partial order and its carrier completely, i.e., use the same symbol for them: $D = (D, \sqsubseteq_D)$ in our example.

In the context of denotational semantics the binary relation $\sqsubseteq_D$ is often considered to be an information approximation, and the binary relation $\sqsubseteq_D$ is called an approximation relation. Similarly, we often say that s_1 approximates s_2 if $s_1 \sqsubseteq s_2$. Next, let us define the notions of monotone functions, upper bounds, least upper bounds, lower bounds and greatest lower bounds.

Definition 2.52 (Monotone Function) *Given two partially ordered sets $(D, \sqsubseteq_D)$ and $(E, \sqsubseteq_E)$. A function $f : D \longrightarrow E$ is monotone if it preserves the partial order, i.e., for all $d_1, d_2 \in D$:*

$$d_1 \sqsubseteq_D d_2 \implies f(d_1) \sqsubseteq_E f(d_2) \tag{2.60}$$

Definition 2.53 (Upper Bound and Least Upper Bound) *Given a partial order $S = (S, \sqsubseteq)$ and a subset $U \subseteq S$. An element $e \in S$ is called an upper bound of U* **iff** *$u \sqsubseteq e$ for all $u \in U$. An element $e \in S$ is called the least upper bound (l.u.b.) of U* **iff** *e is an upper bound of U and for any e' that is an upper bound of U it holds that $e \sqsubseteq e'$. The least upper bound of a set U, which is also called the supremum of U, is unique and is also denoted by $\sqcup_S U$.*

Definition 2.54 (Lower Bound and Greatest Lower Bound) *Given a partial order $S = (S, \sqsubseteq)$ and a subset $U \subseteq S$. An element $e \in S$ is called a lower bound of U* **iff** *$e \sqsubseteq u$ for all $u \in U$. An element $e \in S$ is called the greatest lower bound of U* **iff** *is e is a lower bound of U and for any e' that is a lower bound of U it holds that $e' \sqsubseteq e$. The greatest lower bound of a set U, which is also called the infimum of U, is unique and is also denoted by $\sqcap_S U$.*

Given a partial order $S = (S, \sqsubseteq_S)$ and a subset $U \subseteq S$, we also write $\sqcup U$ and $\sqcap U$ for the least upper bound $\sqcup_S U$ resp. greatest lower bound $\sqcap_S U$ if S is clear from the context. Furthermore, it is usual to talk about $\sqcup U$ as the join of the elements of U and to speak about $\sqcap U$ as the meet of the elements of U. Furthermore, given two elements s and t, it is usual to use the infix notations $s \sqcup t$ and $s \sqcap t$ to denote the least upper bound $\sqcup\{s, t\}$ resp. the greatest lower bound $\sqcap\{s, t\}$ of s and t. Now, a lattice is defined as a partial order in which both the meet and the join exist for all pairs of elements, whereas a complete lattice is a partial order that contains least upper bounds and greatest lower bounds for all of its subsets.

Definition 2.55 (Lattice) *A partial order $(S, \sqsubseteq_S)$ is a lattice* **iff** *there exist the meet $s \sqcap s'$ and the join $s \sqcup s'$ for all elements $s \in S$ and $s' \in S$.*

Definition 2.56 (Complete Lattice) *A partial order $(S, \sqsubseteq_S)$ is a complete lattice* **iff** *there exist the least upper bound $\sqcup U$ and the greatest lower bound $\sqcap U$ for all subsets $U \subseteq S$.*

In particular, given a complete lattice S, we have that both the least upper bound $\sqcup S$ and the greatest lower bound $\sqcap S$ exist and both belong to S. It is usual to call $\sqcap S$ the bottom element of S and to denote it as $\bot_S \in S$ or $\bot \in S$ if S is clear from the context. Similarly, it is usual to call $\sqcup S$ the top element of S and to denote it as $\top_S \in S$ resp. $\top \in S$.

Definition 2.57 (Top and Bottom Elements of Complete Lattices) *Given a complete lattice $(S, \sqsubseteq_S)$ we define*

$$\bot_S = \sqcap S \quad \textit{(bottom element)} \tag{2.61}$$

$$\top_S = \sqcup S \quad \textit{(top element)} \tag{2.62}$$

An important example of complete lattices is the power set $(\mathbb{P}(S), \subseteq)$ of subsets of a given set S ordered by subset inclusion. Another important example is arbitrary closed intervals of real numbers $([r_1, r_2], \leqslant)$ with the usual ordering. With respect to $(\mathbb{P}(S), \subseteq)$ we have that each subset $A \subseteq \mathbb{P}(S)$ has $\cup A$ and $\cap A$ as least upper bound resp. greatest lower bound with respect to subset inclusion. Therefore, $\mathbb{P}(S)$ forms a complete lattice with $\bot = \emptyset$ and $\top = S$. With respect to $([r_1, r_2], \leqslant)$ we have that each $u \subseteq [r_1, r_2]$ has r_1 as lower bound and r_2 as upper bound. Therefore, due to the least-upper-bound property of the real numbers, compare with Axiom 2.88, we have that u also has a supremum $\sqcup u$. The least-upper-bound property is a completeness axiom that ensures that a subset $a \subseteq \mathbb{R}$ has a least upper bound $\sqcup a$ whenever it has an upper bound. Analogously, the greatest-lower-bound property, see Axiom 2.89, ensures the existence of the infimum of u. Therefore, $u \subseteq [r_1, r_2]$ forms a complete lattice with $\bot = r_1$ and $\top = r_2$. However, as a counterexample, the lattice $(\mathbb{R}, \leqslant)$ does not form a complete lattice, but only a so-called condititonal complete lattice, in which the existence of an infimum is only ensured for those subsets that have a lower bound and, analogously, the existence of a supremum is only ensured for subsets that have a upper bound

Now, let us turn to the crucial result of lattice theory, i.e., the Knaster-Tarski theorem, which ensures, among other things, the existence of a least upper bound for every monotone function from a given complete lattice to itself. In [255] the Knaster-Tarski fixed-point theorem is called the lattice-theoretical fixed-point theorem. In an earlier work [160] the Knaster-Tarski theorem was first proven for the special case of the complete lattice $(\mathbb{P}(S), \subseteq)$. Then, in [255] it was generalized to complete lattices.

Theorem 2.58 (Knaster-Tarski Fixed Point Theorem) *Given a complete lattice $A = (A, \sqsubseteq_A)$, and a monotone function $f : A \rightarrow A$, we have that the set $P = \{a \,|\, f(a) = a\}$ of fixed points of f is not empty and the partial order $(P, \sqsubseteq_A)$ is a complete lattice so that the following holds:*

$$\sqcap\{a \,|\, f(a) = a\} = \sqcap\{a \,|\, f(a) \sqsubseteq_A a\} \quad (\bot_P) \tag{2.63}$$

$$\sqcup\{a \,|\, f(a) = a\} = \sqcup\{a \,|\, a \sqsubseteq_A f(a)\} \quad (\top_P) \tag{2.64}$$

Proof. See Theorem 1 and its proof in [255]. □

With respect to inductive definitions we are essentially interested in the least fixed point $\sqcap\{a|f(a) = a\}$ of a given monotone function. Whenever an inductive definition of a set is given it can be made precise by carving out the encompassing domain from which its elements are drawn and defining the monotone function of which it is the least fixed point. Implicitly, we will make use of this several times in this book when we give rule-based definitions, but also explicitly in the definition of the semantic equations that make up the denotational semantics in Chap. 5. Let us have a look at a simple example on how inductive definitions work formally. Let us take the specification of a function as an example. Let us take the familiar factorial function, which is informally defined as follows:

$$f!(n) = 1 \cdot 2 \cdot 3 \cdot \cdots \cdot (n-1) \cdot n \tag{2.65}$$

Now, we define the factorial function as a relation $fac \subseteq \mathbb{N}_0 \times \mathbb{N}_0$. We do so by defining a transformation function t as an endofunction on sets of binary relations over $\mathbb{N}_0$, i.e., a function $t : \mathbb{P}(\mathbb{N}_0 \times \mathbb{N}_0) \longrightarrow \mathbb{P}(\mathbb{N}_0 \times \mathbb{N}_0)$ as follows:

$$t = f \mapsto \{\langle 0, 1\rangle\} \cup \{\langle n+1, (n+1) \cdot i\rangle \,|\, \langle n, i\rangle \in f\} \tag{2.66}$$

Now, it can be shown that $f!$ in Eqn. (2.65) equals the least fixed point of the transformation function t, i.e., we have that $f! = \sqcap\{f \mid t(f) = f\}$. Therefore, it is correct to define *fac* as the least fixed point of t, i.e.,

$$fac = \sqcap\{f \mid t(f) = f\} \tag{2.67}$$

Actually, it can be seen that t is a monotone function with respect to the complete lattice $(\mathbb{P}(\mathbb{N}_0 \times \mathbb{N}_0), \subseteq)$ which guarantees, due to Theorem 2.58, that the least fixed point of t exists. Let us compare the definition of *fac* based on t with the recursive definition of the factorial function in some convenient, usual notation as found in the theory of recursive functions or, up to concrete syntax, found in each high-level programming language:

$$f'(n) = \begin{cases} n \cdot f'(n-1) & , n \geqslant 1 \\ 1 & , n = 0 \end{cases} \tag{2.68}$$

To get the point, first note that the definition of t in Eqn. (2.66) is itself non-recursive. The definition of *fac* becomes inductive only later through its definition as the least fixed point of t in Eqn. (2.67). In particular, although the function variable f in Eqn. (2.66) occurs to the left and to the right of the tuple constructor $\mapsto$, this expresses no recursion in itself, but just the construction of tuples in the space $\mathbb{P}(\mathbb{N}_0 \times \mathbb{N}_0) \times \mathbb{P}(\mathbb{N}_0 \times \mathbb{N}_0)$ that then together form the function t. This is different for the variable f' in Eqn. (2.68). Its occurrence on the left-hand side and right-hand side of the equation directly expresses an inductive resp. recursive definition. However, as a consequence of this, we must pre-assume an appropriate semantics of the notation used in Eqn. (2.68) or must otherwise give an appropriate semantics to it. Now, eventually such a semantics has to be based on some notion of least fixed point. We can give such a semantics directly or we can establish an operational semantics based on some notion of recursive call. However, eventually also an operational semantics has to be based on some inductive definition, e.g., in the definition of an underlying reduction system.

At this point, it is also very important to refer to the theory of recursive functions as established by Kleene [151, 152, 153, 154, 155, 156, 157, 158, 159]. The theory of recursive functions provides a or *the* classical means to give semantics to the notation used in Eqn. (2.68). It does so by reducing it to the well-foundedness of natural numbers. Again, this semantics eventually relies on an inductive definition, i.e., the definition of natural numbers in this case. Now, if you still find it hard to see the correspondence between the definition in Eqn. (2.68) and the definition based on Eqn. (2.66) you might want to rewrite Eqn. (2.66) into the following, completely equivalent form:

$$t = f \mapsto \begin{array}{l} \{\langle n, n \cdot i\rangle \mid n \geqslant 1, \langle n{-}1, i\rangle \in f\} \\ \cup\ \{\langle 0, 1\rangle\} \end{array} \tag{2.69}$$

Next, let us give a further definition of the factorial function that is closer to the convenient notation in Eqn. (2.68) but still completely equivalent to the definitions provided on the basis of t in Eqn. (2.66) or Eqn. (2.69). For this purpose we establish a transformation t' as an endofunction on the partial endofunctions on $\mathbb{N}_0$, i.e., as a function $t' : (\mathbb{N}_0 \nrightarrow \mathbb{N}_0) \longrightarrow (\mathbb{N}_0 \nrightarrow \mathbb{N}_0)$ as follows:

$$t' = f'' \mapsto n \mapsto \begin{cases} n \cdot f''(n-1) & , n \geqslant 1 \\ 1 & , n = 0 \end{cases} \tag{2.70}$$

See, how close Eqn. (2.70) is to the notation in Eqn. (2.68). Now, again it can be shown that $f!$ in Eqn. (2.65) equals and therefore can be defined as the least fixed point of the transformation t'. Actually, given an arbitrary set A we have that the set of partial functions from A to A ordered by subset inclusion, i.e., $(A \nrightarrow A, \subseteq)$, forms what is called a meet-complete semilattice. As in the case of $\mathbb{P}(S)$ for a set S, we have that the greatest lower bound exists for all sets of partial functions $F \subseteq (A \nrightarrow A)$, i.e., $\sqcap F = \cap F$. Therefore, $(A \nrightarrow A, \subseteq)$ forms a meet-complete semi-lattice with the completely undefined function as bottom element, i.e., $\bot = \emptyset$. This already suffices to guarantee the existence of least fixed points of monotone functions and therefore provides a convenient framework to give inductive definitions of sets that draw their elements from A, and all of this in a neat notation as in Eqn. (2.70). Actually, given sets A and B, the set of partial functions $A \nrightarrow B$ can also be naturally embedded into a complete lattice by introducing a top element $\top_B$ to B and defining $f \sqcup f'$ as $(f \sqcup f')(x) = \top_B$ for those x for which both $f(x)$ and $f'(x)$ are defined but $f(x) \neq f'(x)$.

2.4.2 Rule-Based Inductive Definitions

Rule-based definitions provide additional concrete terminology and notation to give inductive definitions. Rule-based definitions can be considered particularly intuitive; in any case, they are extensively used, in particular in computer science, take Martin-Löf type theory [182, 183, 51] as an important example. See also [4] for a standard work on inductive definitions based on rule-based notation. In this book, we will use rule-based notation, e.g., in the definition of type systems of lambda calculi as well as for the one-step semantics of lambda calculi.

Let us give an example first. In rule-based notation, the definition of the factorial function as provided by Eqn. (2.66) can be rewritten as Eqn. (2.71), i.e., the factorial function is defined as the least relation $f \subseteq \mathbb{N}_0 \times \mathbb{N}_0$ that is closed under the following rules:

$$(i)\frac{}{\langle 0, 1 \rangle \in f} \qquad (ii)\frac{\langle n, i \rangle \in f}{\langle n+1, (n+1) \cdot i \rangle \in f} \tag{2.71}$$

Although the definition of the factorial function in Eqn. (2.71) as a least set that is closed under a collection of rules is clear and intuitive, it deserves a little formalization to make it a more rigorous statement. As a first observation, note that Eqn. (2.71) already establishes some convenient notation, e.g., (ii) is actually not a single rule but a rule schema that specifies a set of infinitely many rules, one for each $n \in \mathbb{N}_0$.

Let us start from scratch. We follow [4] to explain rule-based definitions. Basically, a rule is a pair (X, x) of premises $X \subseteq B$ and a conclusion $x \in B$ that draw from the same base set B. Now, a set $A \subseteq B$ is said to be closed under a rule (X, x) if $X \subseteq A$ implies $x \in A$. Furthermore, given a set of rules $\mathcal{R}$ we say that a set $A \subseteq B$ is closed under $\mathcal{R}$ if it is closed under all $r \in \mathcal{R}$. Next, the *set inductively defined by* $\mathcal{R}$ is *defined as* the intersection of all sets A that are closed under $\mathcal{R}$.

Definition 2.59 (Rule-Based Inductive Definitions) *Given a set of rules $\mathcal{R}$ we define the set inductively defined by $\mathcal{R}$, denoted by $I(\mathcal{R})$, as follows:*

$$I(\mathcal{R}) = \bigcap\{A \mid \forall (X, x) \in \mathcal{R} \,.\, X \subseteq A \Rightarrow x \in A\} \tag{2.72}$$

Now, the definition of rule-based induction in Def. 2.59 is obviously well-defined. In Eqn. (2.72), the intersection of all the sets that are closed under $\mathcal{R}$ always exists. However, how do we know that the definition is good? How do we know that it achieves what we wanted, i.e., to yield the least set that is closed under $\mathcal{R}$? Fortunately, to see this, we have the Knaster-Tarski fixed-point theorem 2.58 to hand as well as the discussion in Sect. 2.4.1 on how to base inductive definitions on complete lattices. First, given a set of rules $\mathcal{R}$ so that $X \subseteq B$ and $x \in B$ for all rules $(X, x) \in \mathcal{R}$, let us define the following transformation t as an endofunction on the complete lattice $(\mathbb{P}(B), \subseteq)$ as follows:

$$t = A \mapsto \{x \mid (X, x) \in \mathcal{R},\ X \subseteq A\} \tag{2.73}$$

Now, it can be seen that t is monotone. Therefore, due to the Knaster-Tarski fixed-point theorem 2.58 its least fixed point exists and furthermore equals $\bigcap\{A \mid t(A) \subseteq A\}$. Now, it can be checked that $\bigcap\{A \mid t(A) \subseteq A\}$ equals $I(\mathcal{R})$ as defined in Def. 2.59. Therefore, we have that $\sqcap t$ equals $I(\mathcal{R})$. Furthermore, it can be seen in this way that $I(\mathcal{R})$ is the least set that is closed under $\mathcal{R}$. We see that a rule-based definition is just a means to implicitly specify an appropriate monotone function for which a least fixed point exists.

The usual way to define a rule set is by a set of rule schemes. Conceptually, a single rule scheme somehow has the following form:

$$\frac{P_1 \cdots P_n}{M}\, C_1 \cdots C_m \tag{2.74}$$

A rule scheme lists some assertions P_1 to P_n on set membership to generate the premises of the rules and gives an expression on set membership for generating the conclusion, possibly along with some side conditions C_1 to C_m to narrow the premises. Please have a look at the example in Eqn. (2.71) and how it relates to Eqn. (2.66). Then, several rule schemes are used to further introduce case distinctions. The rules of all rules schemes of a rule-based definition are simply joined.

We deliberately make no attempt here to establish a language of rule schemes and to explain how they exactly specify resp. generate sets of rules. Rather, we give an explicit explanation, which rule sets are meant to be generated or which mathematical objects are involved, whenever we give a rule-based inductive definition and whenever we feel it is needed in the sequel. The point is the following. Rule-based inductive definitions are in wide spread use, in particular in the literature on reduction calculi and the semantics of programming languages. Their meaning is usually obvious as the typical reader can be expected to have enough practice in reading these inductive definitions. Even in theoretical treatments this usually does not pose a problem, because the used mechanism of rule-based definition is considered part of the background mathematics like all the rest of the respective prerequisite mathematical tool kit. In our case, however, it is not satisfying to fully accept inductive definitions as given or granted, because to a large extent inductive definitions are, in their special shape of higher-type computable functions, themselves objects of investigation and clarification. We do not want and in this case actually do not have to rely on the same phenomena of inductive definition in the exploited background mathematics as in the area that we actually want to investigate. Therefore, we will make the background mechanisms at work explicit in the sequel. We use usual rule-based notation in our definitions at several places. However, formalizing the concrete language of these rule-based definitions would be too much overhead. If the reader is nevertheless interested in how a language for rule-based definitions can be fully elaborated we recommend the work [187, 41, 124] on inductive relation definitions with the theorem prover Isabelle/HOL (Higher-Order Logic) [209]. With [187, 41, 124] a concrete package is provided for writing rule-based definitions.

As a minor notational detail, we will also use, e.g., the following usual variants of notation in our rule-based definitions that are meant to denote the same rule set as Eqn. (2.74):

$$\frac{P_1 \ldots P_n \quad C_1 \ldots C_m}{M} \qquad \forall\, C_1 \ldots C_m \,.\, P_1 \ldots P_n \vdash M \tag{2.75}$$

Let us give some further remarks on the notation for rule-based definitions. Most commonly, the set R that is specified is a relation drawn from some space $S_1 \times \cdots \times S_k$. Then, the assertions in rule schemes like Eqn. (2.74) and the conclusion expression are often given in the form $P_i = (P_i' \in R)$ and $M = (M' \in R)$, i.e.,

$$\frac{P_1' \in R \quad \cdots \quad P_n' \in R}{M' \in R}\, C_1 \cdots C_m \tag{2.76}$$

Having a notational form as in Eqn. (2.76), it is then said that the rules define R as the least relation satisfying these rules. This means that R is used as the name of the target object, as well as the name of a relation variable which is bound in the rules. This is similar to inductively defining a set A as the least upper bound of a monotone transformation function $A \mapsto F(A)$, i.e., $A = \sqcap\{A \mid F(A) = A\}$. This style of re-using the name of the target object in the rules is usually also adopted for special relation notation, i.e., it is usual that the notation of the target relation is used in the specification of the rules. As an example, have a look at the specification of the one-step semantics in (3.54) through (3.64). Here, the notation $M \xrightarrow{i} N$ of the target relation is also used in the rules. All of this makes sense and leads to particularly intuitive definitions.

An important use case of rule-based inductive definition, which concerns us in this book, is the specification of term rewriting systems. Given a set of terms $\mathcal{T}$, a term rewriting system establishes a reduction relation $\leadsto \subseteq \mathcal{T} \times \mathcal{T}$ between terms. Therefore if a rule-based induction is used to specify such a reduction relation ρ, both the premises and the conclusions take the form $M \leadsto N$ for some terms M and N. Now, some extra terminology exists for such term tuples $M \leadsto N$. All of these term tuples might be called reductions or also reduction steps. If $M \leadsto N$ is a conclusion, the term M is called a redex, whereas N might itself also be called a reduction, or the reduction of $M \leadsto N$.

2.4.3 Complete Partial Orders and Continuous Functions

Complete partial orders are partially ordered sets in which there exist least upper bounds for all subsets of a certain kind. For example, we have the notion of directed complete partial order, usually abbreviated as dcpo, the notion of chain-complete partial order and the notion of ω-complete partial order.

In a dcpo there exist least upper bounds for all directed subsets. A set is called a directed set it if contains a shared upper bound for all of its pairs of elements. Now, continuous functions for dcpo*s* are functions that preserve least upper bounds of directed subsets. Directed complete partial orders are in widespread use in denotational semantics. Sometimes, they are also called inductive complete partial orders, then abbreviated as ipo*s*. Often, they are just called complete partial orders for short and then abbreviated as cpo*s* instead of dcpo*s*. But note that such cpo*s*, in general, do not enjoy full completeness properties like, e.g., complete lattices. On the contrary, we usually do not use *cpo* to indicate that a partial order contains least upper bounds for all of its subsets. Rather, a partial order that is sure to contain least upper bounds for all of its subsets is called a join-complete semi-lattice.

A chain-complete cpo contains least upper bounds for all of its chains. A set is called a chain if all pairs of its elements are actually related by the respective approximation. An ω-chain is, moreover, a possibly infinite, countable chain. We will base our denotational semantics in Sect. 5 on the notion of ω-cpo*s*. Therefore, we have devoted several sections, i.e., Sects. 2.4.4 through 2.4.8, to the introduction of ω-cpos and the discussion of concepts that are related to them.

2.4.4 ω-Complete Partial Orders

The choice of domains for a concrete denotational semantics depends on the needs of the respective investigation. The choice of ω-chains is one possible, usual choice; see, e.g., [192] and [224].

Definition 2.60 (ω-Chain) *Given a partially ordered set $(S, \sqsubseteq)$, a countable subset of elements $(s_i \in S)_{i\in\omega}$ of S is called an ω-chain in S, or ω-chain for short,* **iff** *$s_i \sqsubseteq s_{i+1}$ for all $i \in \omega$, i.e., $s_0 \sqsubseteq s_1 \sqsubseteq s_2 \sqsubseteq \ldots$.*

Definition 2.61 (Bounds of ω-Chains) *Given a partially ordered set $S = (S, \sqsubseteq)$ and an ω-chain $U = (u_i \in S)_{i\in\omega}$. An element $e \in S$ is called the upper bound of U* **iff** *$u_i \sqsubseteq e$ for all $i \in \omega$. An element $e \in S$ is called the least upper bound of U* **iff** *for any e' that is an upper bound of U it holds that $e \sqsubseteq e'$. The least upper bound of an ω-chain U, which is also called the supremum of U, is unique and is also denoted by $\sqcup_S U$.*

Obviously, an element $e \in S$ is the upper bound of an ω-chain $U : \omega \to S$ if e is an upper bound of the range of U in S, i.e., if e is an upper bound of $U^\dagger(\omega)$. Similarly, the least upper bound of U equals the least upper bound of the range of U, i.e., $\sqcup_S U = \sqcup_S(U^\dagger(\omega))$. Given an ω-chain $(u_i \in S)_{i\in\omega}$ we also use an alternative notation for the least upper bound of such a chain, which gives the ranging index $i \in \omega$ as a subscript to the least upper bound symbol rather than to the chain, i.e., we use all of the following notations equivalently:

$$\sqcup_S(u_i)_{i\in\omega} = \sqcup(u_i)_{i\in\omega} = \bigsqcup_{i\in\omega} u_i = \bigsqcup_{i\in\omega}{}_S\, u_i \tag{2.77}$$

The auxiliary notation can be particularly intuitive when we deal with nested least upper bound structures; see also Lemma 2.75, for example:

$$\bigsqcup_{i\in\omega} \bigsqcup_{j\in\omega} \langle a_i, b_j\rangle = \sqcup(\sqcup(\langle a_i, b_j\rangle)_{j\in\omega})_{i\in\omega} \tag{2.78}$$

Now, let us turn to ω-cpos in Def. 2.62, which play a central role in our denotational semantics of the probabilistic lambda calculus.

Definition 2.62 (ω-Complete Partial Order) *A structure $(S, \sqsubseteq, \perp)$ that consists of a partially ordered set $(S, \sqsubseteq)$ and an element $\perp \in S$ is called an ω-complete partial order, or ω-cpo for short, if every ω-chain $U = (u_i \in S)_{i\in\omega}$ has a least upper bound $\sqcup U \in S$ and $\perp$ is the least element of S with respect to $\sqsubseteq$, i.e., for all $s \in S$ we have that $\perp \sqsubseteq s$.*

2.4.5 ω-CPO Targeting Function Spaces

Next, we are interested in sets of functions that have an ω-cpo as target domain. We will equip these function spaces with a partial ordering and a bottom element and they will turn out to be cpo*s*. Then, we will call them ω-cpo targeting function spaces. In semantics of programming languages we are usually interested in function spaces of ω-continuous functions only; compare with Sect. 2.4.7. An ω-continuous function is a function for which both the source and the target domain are ω-cpo*s*, whereas in case of an ω-cpo targeting function space only the target space is required to be an ω-cpo.

We will use ω-cpo targeting function spaces as base domains in our denotational semantics. For standard, non-probabilistic calculi, you will usually find flat domains as base domains in the denotational semantics. In our semantics, base elements must be functions, because they assign probabilities to even more basic objects, which we will call data points later. However, it will turn out that it is not important that these functions are ω-continuous. It will only be important that these functions form an ω-cpo. We will see that ω-cpo*s* are not needed as source domains to ensure the existence of least upper bounds in function spaces. It suffices that the target domain is an ω-cpo to ensure the general existence of such least upper bounds.

Definition 2.63 (ω-CPO Targeting Function Space) *Given a set S and an ω-cpo $\boldsymbol{E} = (E, \sqsubseteq_E, \bot_E)$ we define the ω-cpo targeting function space $\lfloor S \longrightarrow \boldsymbol{E} \rfloor$ as a structure $(S \longrightarrow E, \sqsubseteq, \bot)$ with $\sqsubseteq \in \mathbb{P}((S \longrightarrow E) \times (S \longrightarrow E))$ and $\bot \in S \longrightarrow E$ so that for all $f_1, f_2 \in S \longrightarrow E$ we have that*

$$f_1 \sqsubseteq f_2 \Leftrightarrow \forall s \in S \,.\, f_1(s) \sqsubseteq_E f_2(s) \tag{2.79}$$

$$\bot = s \in S \mapsto \bot_E \tag{2.80}$$

We immediately have that $(S \longrightarrow E, \sqsubseteq)$ in Def. 2.63 is a partial order. We proceed with proving that the whole structure $\lfloor S \longrightarrow E \rfloor$ forms an ω-cpo. First, as the crucial step, we prove that each ω-chain in $\lfloor S \longrightarrow E \rfloor$ has a least upper bound in Lemma 2.64. Next, the fact that $\lfloor S \longrightarrow E \rfloor$ is an ω-cpo follows immediately from Lemma 2.64 as Corollary 2.65.

Lemma 2.64 (L.U.B.*s* in ω-CPO Targeting Function Spaces) *Given an ω-cpo targeting function space $\lfloor S \longrightarrow E \rfloor$ and an ω-chain $(c_i : S \longrightarrow E)_{i \in \omega}$ the least upper bound $\sqcup_{S \rightarrow E} C$ exists and is given pointwise as follows:*

$$\sqcup C = s \in S \mapsto \sqcup_E \, (c_i(s))_{i \in \omega} \tag{2.81}$$

Proof. As a premise we know that the sequence $(c_i)_{i \in \omega}$ is an ω-chain. Therefore we know that $c_i \sqsubseteq c_{i+1}$ for all $i \in \omega$. This implies, due to the pointwise definition of $\sqsubseteq$ in Eqn. (2.79) that also $c_i(s) \sqsubseteq_E c_{i+1}(s)$ for all $i \in \omega$ and $s \in S$. Therefore, we know that $(c_i(s))_{i \in \omega}$ is an ω-chain for all $s \in S$. Therefore, due to the premise that E is an ω-cpo, we know that the least upper bound

$\sqcup_E (c_i(s))_{i\in\omega}$ exists. Based on that we can define a function $l \in \lfloor S \longrightarrow E \rfloor$ as follows:

$$l = s \in S \mapsto \sqcup_E \ (c_i(s))_{i\in\omega} \tag{2.82}$$

Next, we prove that l is the least upper bound of C, i.e., $l = \sqcup_{S\rightarrow E} C$. This again follows pointwise from the definition of $\sqsubseteq$ in Eqn. (2.79). We know that $c_i(s) \sqsubseteq_E l(s)$ for all $i \in \omega$ and $s \in S$ and therefore $c_i \sqsubseteq l$ for all $i \in \omega$. Furthermore, given a function $l' \in \lfloor S \longrightarrow E \rfloor$ with $c_i \sqsubseteq l'$ for all $i \in \omega$. Then, we know that $c_i(s) \sqsubseteq_E l'(s)$ for all $s \in S$. Therefore, we know that $l(s) \sqsubseteq_E l'(s)$ for all $s \in S$. Therefore we know that $l \sqsubseteq l'$. □

Corollary 2.65 (ω-CPO Targeting Function Spaces are ω-CPOs) *Each ω-cpo targeting function space $\lfloor S \longrightarrow E \rfloor$ is an ω-cpo.*

Proof. With $\perp_{S\rightarrow E}$ there exists a least element in $S \rightarrow E$. The fact that $\perp_{S\rightarrow E}$ is the least element of $S \rightarrow E$ follows pointwise from the definition of $\perp_{S\rightarrow E}$ in Eqn. (2.80). Then, Lemma 2.64 ensures the existence of a least upper bound for each ω-chain in $S \rightarrow E$. □

2.4.6 Product Spaces of ω-CPOs

Before we turn to ω-continuous function spaces in Sect. 2.4.7 we will show how to construct a product space out of two given ω-cpos so that the resulting space is itself an ω-cpo. Consequently, we will call the result of such a construction a product space of ω-cpos. Actually, the construction is straightforward, pointwise, as in the case of the ω-cpo targeting function space seen in Sect. 2.4.5.

Definition 2.66 (Product Space of ω-CPOs) *Given two ω-cpos $\boldsymbol{D} = (D, \sqsubseteq_D, \perp_D)$ and $\boldsymbol{E} = (E, \sqsubseteq_E, \perp_E)$ we define their product space $[\boldsymbol{D} \times \boldsymbol{E}]$ as a structure $(D \times E, \sqsubseteq, \perp)$ with $\sqsubseteq \in \mathbb{P}((D \times E) \times (D \times E))$ and $\perp \in D \times E$ so that for all $d, d' \in D$ and $e, e' \in E$ we have that*

$$\langle d, e \rangle \sqsubseteq \langle d', e' \rangle \iff (d \sqsubseteq_D d' \ \wedge \ e \sqsubseteq_E e') \tag{2.83}$$

$$\perp = \langle \perp_d, \perp_e \rangle \tag{2.84}$$

Lemma 2.67 (Product Spaces of ω-CPOs are ω-CPOs) *Each product space of ω-cpos $[D \times E]$ is an ω-cpo.*

Proof. From the definition of ω-continuous product spaces in Def. 2.66 it follows component-wise that $\langle \perp_d, \perp_e \rangle$ is the least element of $[D \times E]$ and, furthermore, that for each chain $(\langle d_i, e_i \rangle)_{i\in\omega}$ there exists a least upper bound in $[D \times E]$. □

Product spaces of ω-cpo*s* are not needed directly in the definition of our denotational semantics; nevertheless, they are important to consider. The probabilistic lambda calculus that we will consider is reductionist with respect to its types. It knows only ground types and functional types. As usual, due to currying [236], product types do not add a conceptual value to the investigation. This means that we have no product types in our calculus and therefore we also do not need product spaces of ω-cpo*s* as domains in the definition of our denotational semantics. However, we technically need them, because they emerge implicitly, on the fly, in some crucial proofs in the book. For example, we need them in proving that the semantic fixed-point operator Φ is ω-continuous; see Lemma 2.82. Also, we need them to prove that semantic update preserves ω-continuity; see Lemma 5.21. In both cases we turn a function application into an explicit apply operator; compare with Def. 2.73, which causes product domains to appear.

2.4.7 ω-Continuous Function Spaces

The notion of ω-continuous function plays a central role. An ω-continuous function is a function between two ω-cpo*s* that is monotone and, moreover, preserves least upper bounds of ω-chains. These properties will be exploited later, to prove the existence of fixed points; compare with Theorem 2.79. First, we introduce images of ω-cpo*s*. Then we define ω-continuous functions and, in particular, ω-continuous function spaces. After a series of important technical lemmas we will prove that ω-continuous function spaces are themselves cpo*s*.

Definition 2.68 (Images of ω-Chains) *Given two sets A and B, a function $f : A \to B$ and an ω-chain $\boldsymbol{a} = (a_i \in A)_{i \in I}$. We define the image of $\boldsymbol{a}$ under f, also denoted by $f^\dagger(\boldsymbol{a})$, as an indexed family as follows:*

$$f^\dagger(\boldsymbol{a}) = (f(a_i))_{i \in I} \tag{2.85}$$

Note that we overload the notation $f^\dagger((a_i \in A)_{i \in I})$ for the image of an ω-chain with the notation for the image $f^\dagger(A)$ of some set A under the function f. Technically, we often need the fact that the image of an ω-chain is again an ω-chain, as expressed by Lemma 2.69.

Lemma 2.69 (Images of ω-Chains) *Given two ω-cpos $(A, \sqsubseteq, \bot)$ and $(B, \sqsubseteq, \bot)$ and a monotone function $f : A \to B$. The image of an ω-chain $(a_i \in A)_{i \in \omega}$ under f, i.e., $(f(a_i) \in B)_{i \in \omega}$, is again an ω-chain.*

Proof. This follows immediately by the fact that f is a monotone function. We have that $a_i \sqsubseteq a_{i+1}$ for all $i \in \omega$, because $(a_i \in A)_{i \in \omega}$ is an ω-chain. Therefore, and due to the monotonicity of f, we have that $f(a_i) \sqsubseteq f(a_{i+1})$ for all $i \in \omega$. □

Next, an ω-continuous function is a monotone function between ω-cpo*s* that, moreover, preserves least upper bounds. We will take the set of ω-continuous functions between two ω-cpo*s* and turn it into a structure which is called an ω-continuous function space. Then, we will be able to show that ω-continuous function spaces are ω-cpo*s*. Given two ω-cpo*s* E and D, we use $[E \longrightarrow D]$ to denote both the set of ω-continuous functions between E and D as well as the ω-continuous function space constructed for E and D.

Definition 2.70 (ω-Continuous Function) *Given two ω-cpos $D = (D, \sqsubseteq_D, \bot_D)$ and $E = (E, \sqsubseteq_E, \bot_E)$. A function $f : D \longrightarrow E$ is ω-continuous* **iff** *it is monotone and preserves least upper bounds of ω-chains, i.e., for all ω-chains $U = (u_i \in D)_{i\in\omega}$ it holds that*

$$f(\sqcup_D U) = \sqcup_E \, (f(u_i))_{i\in\omega} \tag{2.86}$$

Definition 2.71 (Set of ω-Continuous Functions) *Given two ω-cpos $D = (D, \sqsubseteq_D, \bot_D)$ and $E = (E, \sqsubseteq_E, \bot_E)$, we define the set of ω-continuous functions between E and D, denoted by $[D \longrightarrow E]$, as follows:*

$$[D \longrightarrow E] = \{f : D \longrightarrow E \mid f \; is \; \omega - continuous\} \tag{2.87}$$

Definition 2.72 (ω-Continuous Function Space) *Given two ω-cpos $D = (D, \sqsubseteq_D, \bot_D)$ and $E = (E, \sqsubseteq_E, \bot_E)$, we define the ω-continuous function space $[D \longrightarrow E]$ as the structure $([D \longrightarrow E], \sqsubseteq, \bot)$ with approximation relation $\sqsubseteq \in \mathbb{P}([D \longrightarrow E] \times [D \longrightarrow E])$ and bottom element $\bot \in [D \longrightarrow E]$ so that for all $f_1, f_2 \in [D \longrightarrow E]$ we have that*

$$f_1 \sqsubseteq f_2 \Leftrightarrow \forall d \in D \, . \, f_1(d) \sqsubseteq_E f_2(d) \tag{2.88}$$

$$\bot = d \in D \mapsto \bot_E \tag{2.89}$$

Note that the definition of ω-continuous function in Def. 2.70 is well defined, i.e., the sequence $(f(u_i))_{i\in\omega}$ needed to define ω-continuity is actually an ω-chain, because the function f is required to be monotone by Def. 2.70; compare with Lemma 2.69. Therefore, we know that also the least upper bound of $(f(u_i))_{i\in\omega}$ actually exists.

Next, we make explicit function application as a family of *apply* operators to ease argumentation in the sequel. We can perceive a function application $f(d)$ also as the result of applying a semantic application operator at one level higher in the functional hierarchy. The operator, which we call *apply*, is a function $apply : (((A \rightarrow B) \times A) \rightarrow B)$ for all sets A and B, i.e., it takes a function $f : A \rightarrow B$ and data $d : A$ as its arguments and yields the value of applying f to d as its result. This viewpoint is exactly the viewpoint established by c.c.c. models (Cartesian closed category) [173, 144] of typed lambda calculi. As an important technical lemma, we will prove that each apply operator is monotone in both of its arguments, if these arguments are ω-cpo*s*; see Lemma 2.74.

Definition 2.73 (The Apply Operator) *For each pair of sets A and B the apply operator is defined for all $f : A \rightarrow B$ and $d : A$ as the following function:*

$$apply : ((A \longrightarrow B) \times A) \longrightarrow B \tag{2.90}$$

$$apply(f, d) = f(d) \tag{2.91}$$

Lemma 2.74 (Monotonicity of Function Application) *Given ω-cpos $A = (A, \sqsubseteq, \bot)$ and $B = (B, \sqsubseteq, \bot)$, the apply operator for the ω-continuous function space $[[[A \to B] \times A] \to B]$ is monotone in both of its arguments, i.e., for all $f, f' : [A \to B]$ and $d, d' \in A$ we have that*

$$f \sqsubseteq f' \implies apply(f, d) \sqsubseteq apply(f', d) \tag{2.92}$$

$$d \sqsubseteq d' \implies apply(f, d) \sqsubseteq apply(f, d') \tag{2.93}$$

Proof. First, with respect to Eqn. (2.92) we have that $f \sqsubseteq f'$ implies $f(d) \sqsubseteq f'(d)$, which follows pointwise. Next, with respect to Eqn. (2.93) we have that $d \sqsubseteq d'$ implies $f(d) \sqsubseteq f(d')$, which follows by the fact that $f \in [A \to B]$ is ω-continuous and therefore monotone. Now, the Lemma follows immediately by the definition of the apply operator in Def. 2.73. □

Next, we will show that each ω-continuous function space is an ω-cpo. The proof of this proposition, see Corollary 2.78, is a refinement of the proof of Corollary 2.65 in which we have shown that each ω-cpo targeting function space is an ω-cpo. The basic steps concerning the existence of a bottom element and the existence of a least upper bound for each chain in the function space remain exactly the same. The difference is that we cannot be sure any more that the least upper bound that we construct for a chain of functions is itself an ω-continuous function. However, we want it to be an ω-continuous function, otherwise it would lay outside the defined ω-continuous function space.

You can find a proof that ω-continuous function spaces are ω-cpos also as Proposition 2.1 in [192]. In typical proofs in textbooks or, e.g., the proof in [192], the proofs of a l.u.b. exchange lemma and a proof about least upper bounds of ω-continuous function chains are typically merged together into one proof about ω-continuous function spaces. We have turned them into a series of explicit lemmas, because we will need them later in proofs, e.g., in the proof that our denotational semantics is well defined.

Lemma 2.75 (L.U.B. Elimination Lemma) *Given two ω-cpos A and B and a function $f : A \times B \to D$ that is monotone in both of its arguments. Then, for all chains $a = (a_i \in A)_{i\in\omega}$ and $b = (b_i \in A)_{i\in\omega}$ we have that*

$$\bigsqcup_{m\in\omega} \big(\bigsqcup_{n\in\omega} f(a_m, b_n) \big) = \bigsqcup_{i\in\omega} f(a_i, b_i) \tag{2.94}$$

$$\bigsqcup_{n\in\omega} \big(\bigsqcup_{m\in\omega} f(a_m, b_n) \big) = \bigsqcup_{i\in\omega} f(a_i, b_i) \tag{2.95}$$

Proof. We show Eqn. (2.94) only. First it is necessary to see that all involved least upper bounds actually exist. The fact that A and B are cpos, the fact that f is monotone and Lemma 2.69 ensure the existence of all of the following sequences together with the respective least upper bounds: the sequence $f(a_i,b_i)_{i\in\omega}$, all sequences of the form $f(a_m,b_n)_{n\in\omega}$ for arbitrary but fixed a_m as well as the sequence $(\sqcup f(a_m,b_n)_{n\in\omega})_{m\in\omega}$.

We start with proving that $\sqcup(\sqcup f(a_m,b_n)_{n\in\omega})_{m\in\omega} \sqsubseteq \sqcup f(a_i,b_i)_{i\in\omega}$. Consider an arbitrary element $f(a_m,b_n) \in D$ for some $m,n \in \omega$. Without loss of generality, we can assume that $m \leqslant n$. Now, we know that $a_m \sqsubseteq a_n$, because a is an ω-chain. Therefore, we know due to the monotonicity of f in both of its arguments that $f(a_m,b_n) \sqsubseteq f(a_n,b_n)$. Now, we know that $f(a_n,b_n)$ is smaller than the least upper bound of $f(a_i,b_i)_{i\in\omega}$. Altogether we have that

$$\forall m \in \omega . \forall n \in \omega . f(a_m,b_n) \sqsubseteq \sqcup f(a_i,b_i)_{i\in\omega} \tag{2.96}$$

Let us take a_m as arbitrary but fixed. Due to Eqn. (2.96) we know that $\sqcup f(a_i,b_i)_{i\in\omega}$ is an upper bound of the ω-chain $f(a_m,b_n)_{n\in\omega}$ and therefore also $\sqcup f(a_m,b_n)_{n\in\omega} \sqsubseteq \sqcup f(a_i,b_i)_{i\in\omega}$. But from the latter it follows that $\sqcup f(a_i,b_i)_{i\in\omega}$ is also an upper bound of the ω-chain $(\sqcup f(a_m,b_n)_{n\in\omega})_{m\in\omega}$ which amounts to $\sqcup(\sqcup f(a_m,b_n)_{n\in\omega})_{m\in\omega} \sqsubseteq \sqcup f(a_i,b_i)_{i\in\omega}$.

Next, we show that $\sqcup f(a_i,b_i)_{i\in\omega} \sqsubseteq \sqcup(\sqcup f(a_m,b_n)_{n\in\omega})_{m\in\omega}$. For each arbitrary but fixed $j \in \omega$, we know that $f(a_j,b_j)$ must be smaller than the least upper bound of $f(a_i,b_i)_{i\in\omega}$, i.e., $f(a_j,b_j) \sqsubseteq \sqcup f(a_i,b_i)_{i\in\omega}$. Furthermore, we can see that $f(a_j,b_j) \sqsubseteq \sqcup(\sqcup f(a_m,b_n)_{n\geqslant j})_{m\geqslant j}$. Therefore we also know that $f(a_j,b_j) \sqsubseteq \sqcup(\sqcup f(a_m,b_n)_{n\in\omega})_{m\in\omega}$, i.e., $\sqcup(\sqcup f(a_m,b_n)_{n\in\omega})_{m\in\omega}$ is an upper bound of $f(a_j,b_j)$ and therefore $\sqcup f(a_i,b_i)_{i\in\omega} \sqsubseteq \sqcup(\sqcup f(a_m,b_n)_{n\in\omega})_{m\in\omega}$. □

Corollary 2.76 (L.U.B. Exchange Lemma) *Given two ω-cpos A and B and a function $f : A \times B \to D$ that is monotone in both of its arguments. Then, for all chains $(a_i \in A)_{i\in\omega}$ and $(b_i \in A)_{i\in\omega}$ we have that*

$$\bigsqcup_{m\in\omega}\Big(\bigsqcup_{n\in\omega} f(a_m,b_n)\Big) = \bigsqcup_{n\in\omega}\Big(\bigsqcup_{m\in\omega} f(a_m,b_n)\Big) \tag{2.97}$$

Proof. Immediate corollary from the two parts of the l.u.b. elimination Lemma, i.e., Lemma 2.75, as follows:

$$\bigsqcup_{m\in\omega}\Big(\bigsqcup_{n\in\omega} f(a_m,b_n)\Big) = \bigsqcup_{i\in\omega} f(a_i,b_i) = \bigsqcup_{n\in\omega}\Big(\bigsqcup_{m\in\omega} f(a_m,b_n)\Big)$$

□

Lemma 2.77 (ω-Continuity of Least Upper Bounds) *Given an ω-continuous function space $[D \to E]$ and an ω-chain F of functions in $[D \to E]$, then the least upper bound $\sqcup F$ is an ω-continuous function.*

Proof. We assume that F is given as $F = (f_i \in [D \to E])_{i\in\omega}$. According to Def. 2.70, we need to show for all chains $D' = (d_i \in D)_{i\in\omega}$ we have that

$$(\sqcup F)(\sqcup D') = \sqcup(\ (\sqcup F)(d_i)\ )_{i \in \omega} \tag{2.98}$$

Due to Lemma 2.64 we know that the least upper bound of F is given pointwise so that the left-hand side of Eqn. (2.98) equals the following:

$$\bigsqcup_{i \in \omega} (\ f_i(\sqcup D')\ ) \tag{2.99}$$

Next, we can exploit that all f_i in the ω-chain F are ω-continuous functions, so that Eqn. (2.99) equals

$$\bigsqcup_{i \in \omega} (\ \bigsqcup_{j \in \omega} f_i(d_j)\ ) \tag{2.100}$$

Next, the application function in Eqn. (2.100) can be made explicit yielding the following formula:

$$\bigsqcup_{i \in \omega} (\ \bigsqcup_{j \in \omega} apply(f_i, d_j)\ ) \tag{2.101}$$

Now, we know due to Lemma 2.74 that *apply* is monotone in both of its arguments and therefore it is possible to apply the l.u.b. exchange Lemma 2.76 to Eqn. (2.101) so that Eqn. (2.101) equals

$$\bigsqcup_{j \in \omega} (\ \bigsqcup_{i \in \omega} apply(f_i, d_j)\ ) = \bigsqcup_{j \in \omega} (\ \bigsqcup_{i \in \omega} f_i(d_j)\ ) \tag{2.102}$$

Now, again by the pointwise definition of $\sqcup F$, i.e., due to Lemma 2.64, we know that Eqn. (2.102) equals

$$\sqcup(\ (\sqcup F)(d_j)\ )_{j \in \omega} \tag{2.103}$$

□

Corollary 2.78 (ω-Continuous Function Spaces are ω-CPOs)
Each ω-continuous function space $[D \longrightarrow E]$ is an ω-cpo.

Proof. The pointwise definition of $\bot_{S \to E}$ in Eqn. (2.89) ensures that $\bot_{S \to E}$ is the least element of $[D \longrightarrow E]$. Next, due to Lemma 2.64, we know that for each ω-chain F in $[D \longrightarrow E]$ there exists a least upper bound $\sqcup_{S \to E} F$ in $S \to E$. It remains to show that $\sqcup_{S \to E} F$ also is an element of $[D \longrightarrow E]$, i.e., that $\sqcup_{S \to E} F$ is ω-continuous. But this is exactly what has been proven as Lemma 2.77. □

2.4.8 Fixed Points in ω-CPOs

In this section we prove the fixed-point theorem for ω-cpos, which ensures the existence of a least fixed point for ω-continuous functions of the form $f \in [D \longrightarrow D]$. The fixed-point theorem is crucial for the development of the denotational semantics. We will turn the uniquely existing fixed point into an operator, i.e., we define a family of operators $\Phi : (D \to D) \to D$ for each ω-cpo D that yields the fixed point for each argument function. Then, the fixed-point operator Φ will be exploited in the denotational semantics of Chap. 5 to define the semantics of recursion in our probabilistic lambda calculus. Furthermore, we need to prove monotonicity and ω-continuity of the fixed-point operator, in order to prove the l.u.b. substitution Lemma 5.21 later, which is the basis for proving the denotational semantics of lambda abstractions to be well defined.

Theorem 2.79 (Fixed-Point Theorem for ω-CPOs) *Given an ω-cpo D and an ω-continuous function $f \in [D \longrightarrow D]$. Then, there exists a least fixed point $d \in D$ of f and, moreover, d can be constructed as the least upper bound of an ω-chain in D, i.e.,*

1. $f(d) = d$
2. *for all d' with $f(d') = d'$ it holds that $d \sqsubseteq_D d'$*
3. $d = \bigsqcup_{n \geqslant 0} f^n(\bot_D)$

Proof. See, e.g., Proposition 2.2 in [192] or Theorem 4.12 in [118]. First, we need to show that $(f^n(\bot_D))_{n\in\omega}$ is actually an ω-chain. But the ω-chain property $f^n(\bot_D) \sqsubseteq_D f^{n+1}(\bot_D)$ for all $n \in \mathbb{N}_0$ can immediately be seen by a natural induction on n. First, due to the minimality of $\bot_D$ and $f^0(\bot_D) = \bot_D$ we have that $f^0(\bot_D) \sqsubseteq f^1(\bot_D)$ as induction anchor. Next, we know that f is monotone, because f is ω-continuous. Therefore, as induction step for $n \geqslant 1$, we know that $f^{n-1}(\bot_D) \sqsubseteq f^n(\bot_D)$ implies $f(f^{n-1}(\bot_D)) \sqsubseteq f(f^n(\bot_D))$, i.e., $f^n(\bot_D) \sqsubseteq f^{n+1}(\bot_D)$. Next, we show that $\bigsqcup_{n\geqslant 0} f^n(\bot_D)$ is a fixed point of f. Due to the ω-continuity of f we know that $f(\bigsqcup_{n\geqslant 0} f^n(\bot_D))$ equals $\bigsqcup_{n\geqslant 0}(f(f^n(\bot_D)))$ which can be written as $\bigsqcup_{n\geqslant 0} f^{n+1}(\bot_D)$ and equally well as $\bigsqcup_{n\geqslant 1} f^n(\bot_D)$. Now, again due to $f^0(\bot_D) \sqsubseteq f^1(\bot_D)$ we have that $\bigsqcup_{n\geqslant 1} f^n(\bot_D)$ equals $\bigsqcup_{n\geqslant 0} f^n(\bot_D)$. Next, we show the minimality of $\bigsqcup_{n\geqslant 0} f^n(\bot_D)$. Assume that we have an element $d' \in D$ so that $f(d') = d'$. Now, due to the minimality of $\bot_D$ we know that $\bot_D \sqsubseteq d'$. Now, due to the monotonicity of f it can be shown by natural induction that $f^n(\bot) \sqsubseteq f^n(d') = d'$ for all n. This means that d' is an upper bound of $(f^n(\bot))_{n\geqslant 0}$ and therefore we also know that $\bigsqcup_{n\geqslant 0} f^n(\bot_D) \sqsubseteq d'$.

□

Definition 2.80 (Fixed-Point Operator) *Given an ω-cpo D, the fixed-point operator $\Phi : [[D \longrightarrow D] \longrightarrow D]$, which yields, according to Theorem 2.79, for every function $f : [D \longrightarrow D]$ the least fixed point of f, is defined as follows:*

$$\Phi(f) = \bigsqcup_{n\geqslant 0} f^n(\bot_D) \tag{2.104}$$

Corollary 2.81 (Monotonicity of the Fixed-Point Operator) *Given an ω-cpo D, the fixed-point operator Φ is monotone.*

Proof. Direct Corollary from the definition of the fixed-point operator and the fixed-point Theorem 2.79. Given functions $f_1 \sqsubseteq f_2$ we have that $\Phi f_1 = f_1$ and $\Phi f_2 = f_2$ and therefore $\Phi f_1 \sqsubseteq \Phi f_2$.

Lemma 2.82 (ω-Continuity of the Fixed-Point Operator) *Given an ω-cpo D, the fixed-point operator Φ is ω-continuous.*

Proof. See, e.g., Proposition 3 in [192]. Due to Corollary 2.81 we already know that the fixed-point operator Φ is monotone. Therefore it remains to show that for all ω-chains $F = (f_i \in [D \to D])_{i \in \omega}$ we have that

$$\Phi(\sqcup F) = \sqcup(\Phi^\dagger F) \tag{2.105}$$

Let us write Eqn. (2.105) more explicitly in terms of the constituting chain elements, i.e.,

$$\Phi(\bigsqcup_{i \in \omega}(f_i)) = \bigsqcup_{i \in \omega}(\Phi f_i) \tag{2.106}$$

We show that both $\Phi(\bigsqcup_{i \in \omega}(f_i)) \sqsubseteq \bigsqcup_{i \in \omega}(\Phi f_i)$ and $\bigsqcup_{i \in \omega}(\Phi f_i) \sqsubseteq \Phi(\bigsqcup_{i \in \omega}(f_i))$ and start with the latter. We know that $f_i \sqsubseteq \bigsqcup_{i \in \omega}(f_i)$ for all $i \in \omega$. Then, we know due to the monotonicity of the fixed-point operator, i.e., Corollary 2.81, that $\Phi f_i \sqsubseteq \Phi(\bigsqcup_{i \in \omega}(f_i))$ for all $i \in \omega$, which means nothing else but $\Phi(\bigsqcup_{i \in \omega}(f_i))$ is an upper bound for all Φf_i and therefore $\bigsqcup_{i \in \omega}(\Phi f_i) \sqsubseteq \Phi(\bigsqcup_{i \in \omega}(f_i))$.

We proceed to show that $\Phi(\bigsqcup_{i \in \omega}(f_i)) \sqsubseteq \bigsqcup_{i \in \omega}(\Phi f_i)$. Since we know that $\Phi(\bigsqcup_{i \in \omega}(f_i))$ is the least fixed point of $\sqcup F$ it suffices to show that $\bigsqcup_{i \in \omega}(\Phi f_i)$ is also a fixed point of $\sqcup F$ in order to prove this. Now, the target is to prove that $(\sqcup F)(\bigsqcup_{i \in \omega}(\Phi f_i)) = \bigsqcup_{i \in \omega}(\Phi f_i)$. Now, due to the definition of the least upper bound in Lemma 2.64 we have that $(\sqcup F)(\bigsqcup_{i \in \omega}(\Phi f_i))$ equals

$$\bigsqcup_{j \in \omega}(\, f_j\, (\bigsqcup_{i \in \omega}(\Phi f_i))\,) \tag{2.107}$$

Now we know that each $f_j \in [D \to D]$, i.e., that each f_j is ω-continuous. Therefore, we have that Eqn. (2.107) equals

$$\bigsqcup_{j \in \omega}(\,\bigsqcup_{i \in \omega}(f_j(\Phi f_i))\,) \tag{2.108}$$

Now we can introduce the explicit *apply* operator from Def. 2.73 to Eqn. (2.108) so that it is turned into the following formula:

$$\bigsqcup_{j \in \omega}(\,\bigsqcup_{i \in \omega} apply(f_j, \Phi f_i)\,) \tag{2.109}$$

Given the monotonicity of the *apply* operator as stated in Lemma 2.74, it is possible to apply the l.u.b. elimination Lemma 2.75 to Eqn. (2.109) so that Eqn. (2.109) is shown to equal

$$\bigsqcup_{i\in\omega} apply(f_i, \Phi f_i) = \bigsqcup_{i\in\omega} (f_i(\Phi f_i)) \tag{2.110}$$

Finally, due to the fact that Φf_i is a fixed point for all of the f_i we have that Eqn. (2.110) equals

$$\bigsqcup_{i\in\omega} (\Phi f_i) \tag{2.111}$$

□

2.5 Miscellaneous

The purpose of this section is to compile further, rather basic mathematical concepts and notation that are used in the book. We have selected just a few concepts, for which we felt that it is convenient to present them at a single spot. We do not discuss concepts from automata theory here, albeit these concepts are very important for this book, i.e., in the treatment of termination behavior in Chap. 4. We take all the important notions, such as recursive languages, recursive functions, total recursive functions etc. for granted. As usual, it is necessary to mention [137] as an authoritative standard textbook with respect to automata theory and formal languages. In particular, the fundamental notions of formal languages are important for us and, also, knowledge about how they are practically exploited in the specification of languages and calculi. For example, a robust understanding of abstract syntax versus concrete syntax as found in the field of compiler construction is important. As usual we refer to the well-known "dragon book" [5] as reference for compiler construction material.

2.5.1 Set and Function Notation

In this section we compile and fix some important terminology and notation for sets and functions that is used throughout the book.

Given sets S and T, we denote the complement of T in S, which is defined as $\{s \in S \mid s \notin T\}$, by $S \backslash T$ or $S - T$. If we can assume S is known from the context and $T \subseteq S$ we denote $S \backslash T$ also by $\overline{T}$, which is then just called the complement of T for short. Given sets A, B and U such that $A \subseteq U$ and $B \subseteq U$, we have that $A \backslash B$ equals $A \cap \overline{B}$, where $\overline{B}$ is assumed to denote the complement of B in U. Given sets A, B and C we have that $A \backslash (B \cap C)$ equals $A \backslash B \cup A \backslash C$. We denote the one-element set by $\mathbf{1}$ and its element by $\underline{1}$, i.e., $\mathbf{1} = \{\underline{1}\}$.

Given a function $f : S \longrightarrow T$. We use $f^{-1} : T \longrightarrow \mathbb{P}(S)$ to denote the inverse function of f which is defined by $f^{-1}(t) =_{DEF} \{s \in S \mid f(s) = t\}$. We call $f^{-1}(t)$ the inverse image of t under f. We use $f^{\dagger} : \mathbb{P}(S) \longrightarrow \mathbb{P}(T)$ to denote the lifting of f to sets of source and target elements which is defined for all subsets $U \subseteq S$ as $f^{\dagger}(U) =_{DEF} \{f(u) \mid u \in U\}$. We say that $f^{\dagger}(U)$

is the image of U under f, or simply f's image of U. We call $f^\dagger(S) \subseteq T$ the range of the function f. The set S is usually called the domain of the function, whereas T is called the codomain or also the range of the function. This means that the use of range of a function is ambiguous. Furthermore, with domain we usually refer to more complex structures or spaces than sets, i.e., the structures that denotational semantics is based on. All of these issues are mere presentational issues, because it should always be clear from the context which kind of concrete mathematical structure is meant in each case.

Given a function $f : S \to T$ with $f(d) = F(d)$ for each $d \in S$ for some expression $F(d)$, we use the notation $d \mapsto F(d)$ to denote the function f, i.e., $f = d \mapsto F(d)$. An expression of the form $d \mapsto F(d)$ is called a semantic abstraction. The semantic abstraction notation can be used to avoid the introduction of additional, intermediate function symbols. For example, we can write $((d \mapsto +1(d))x)$ instead of writing $f(x), where\ f(d) = +(1(d))$. In semantic equations it is used to make the definition of functions more explicit, i.e., we often write $f =_{DEF} d \mapsto F(d)$ instead of writing $f(d) =_{DEF} F(d)$.

We denote the partial function space between sets A and B by $A \twoheadrightarrow B$. We denote the set of injective functions, also called embeddings, between sets A and B by $A \hookrightarrow B$. A function $f : S \longrightarrow S'$ is called an endofunction if and only if $S = S'$. Given a set S, the identity function on S is denoted by $\mathbf{id}_S : S \to S$ and is defined as $x \in S \mapsto x$. If S is clear from the context, $\mathbf{id}_S$ is also written **id** for short.

Definition 2.83 (n-fold Function Application) *Given an endofunction $f : S \longrightarrow S$ and an object $d \in S$, we define $f^n(d)$, i.e., the n-fold application of f to d, for each $n \geqslant 0$ as follows:*

$$f^n(d) = \begin{cases} d & , n = 0 \\ f(f^{n-1}(d)) & , else \end{cases} \tag{2.112}$$

Definition 2.84 (Function Update) *Given a function $\phi : S \longrightarrow T$, an element $x \in S$ and an element $d \in T$, we define the update of the function ϕ with respect to the argument x (at the point x) by data d, denoted by $\phi[x := d]$, for each $s \in S$ as follows:*

$$\phi[x := d](s) = \begin{cases} d & , s = x \\ \phi(s) & , else \end{cases} \tag{2.113}$$

2.5.2 Indexed Families and Sequences

Indexed families are a basic mathematical notation. Given a set S, an indexed family $e = (e_i)_{i \in I}$ of elements of S indexed by I is a function $e : I \longrightarrow S$ so that $e_i = e(i)$ for all $i \in I$. Then, the set I is called the index set and S is said to be indexed by I. The function $e : I \longrightarrow S$ is just a means to list and access objects in S via I. Sequences are sets indexed by the natural numbers or starting fragments of the natural numbers. Often, counting of sequence elements may start with zero, as is usual for walks and paths in graphs; compare with Defs. 2.35 and 2.40. Also often, counting of sequence elements may start with one, as is usual for n-vectors; see Sect. 2.5.5. A finite sequence $e = (e_i)_{i\in\{0,\ldots,n\}}$ is also written $e_0e_1e_2\cdots e_n$. We denote its last element e_n by $[e]^{\blacktriangleright}$, i.e., $[e_0e_1e_2\cdots e_n]^{\blacktriangleright} = e_n$. Given a sequence e we denote its length by $\#(e)$. Similarly, given a set of sequences E we denote the maximum length of sequences in E as $\#(E)$. Given a finite sequence s and a finite or infinite sequence t we denote the concatenation of s and t by $s \bullet t$.

Next, we recap the usual Kleene closure notation that provides neat notation for certain sets of sequences. Given a set S, its empty sequence ϵ, the set S^n of its element sequences of length n, the set $S^\star$ of its element sequences of finite length, and the set S^+ of its non-empty element sequences of finite length are defined as follows:

$$S^0 = \{\epsilon\}\ \text{, with arbitrary but fixed } \epsilon \notin S \tag{2.114}$$

$$S^n = \{(s_i)_{i\in\{0,..,n-1\}} \mid s_i \in S\ \}\ , n \geqslant 1 \tag{2.115}$$

$$S^\star = \{s \mid n \geqslant 0, s \in S^n\ \} \tag{2.116}$$

$$S^+ = \{s \mid n \geqslant 1, s \in S^n\ \} \tag{2.117}$$

Note that the set $\{\epsilon\}$ is a one-element set for an arbitrarily chosen but fixed object $\epsilon \notin S$ not yet contained in S, which represents the empty sequence of elements of S. The set $S^\star$ in Eqn. (2.116) is called the Kleene closure of S, or also the Kleene hull of S.

2.5.3 Equivalence Relations

In this section we recap the notions of equivalence relation, equivalence class and quotient set. We need these concepts in Sect. 4.3.2 in order to explain the notion of program run event, which is an equivalence class of program executions that cannot be distinguished further by the underlying Markov chain semantics; compare with Def. 4.12.

Definition 2.85 (Equivalence Relation) *A binary relation* $\approx: S \times S$ *on* S *is an equivalence realation if it is reflexive, symmetric and transitive, i.e., for all* $s, s_1, s_2, s_3 \in S$ *it holds that*

1. $s \approx s$ *(reflexivity)*
2. $s_1 \approx s_2 \implies s_2 \approx s_1$ *(symmetry)*
3. $s_1 \approx s_2 \wedge s_2 \approx s_3 \implies s_1 \approx s_3$ *(transitivity)*

Definition 2.86 (Equivalence Class of an Equivalence Relation) *Given an equivalence relation* $\approx: S \times S$ *and an element* $e \in S$ *we define the equivalence class of* e *with respect to* $\approx$, *which is denoted by* $[e]_{/\approx}$, *as follows:*

$$[e]_{/\approx} = \{e' \mid e' \approx e\} \tag{2.118}$$

Definition 2.87 (Quotient Set) *Given an equivalence relation* $\approx: S \times S$ *we define the quotient set* $S_{/\approx}$ *of* S *under* $\approx$ *as follows:*

$$S_{/\approx} = \{[s]_{/\approx} \mid s \in S\} \tag{2.119}$$

2.5.4 Real Analysis

Real analysis is important for us, because the domains of our semantics are based on real numbers. The foundational axioms of real analysis, which will be exploited in semantic argumentations, are equally important for us as basic terminology and notation from real analysis. For thorough treatments of real analysis, please have a look, e.g., at [6, 234].

Axiom 2.88 (Least-Upper-Bound Property) *If a set of real numbers has an upper bound then it has a least upper bound.*

Axiom 2.89 (Greatest-Lower-Bound Property) *If a set of real numbers has a lower bound then it has a greatest lower bound.*

Definition 2.90 (Convergent Sequences and Limits) *A sequence of real numbers* $(a_i)_{i\in\omega}$ *is called convergent* **iff** *there exists an* $a \in \mathbb{R}$ *such that for all* $\epsilon \in \mathbb{R}$ *there exists an* $n \in \mathbb{N}_0$ *such that for all* $n' > n$ *we have that* $|a - a_{n'}| < \epsilon$*. Then, the number* a *is called the limit of* $(a_i)_{i\in\omega}$ *and is denoted by* $\lim_{n\to\infty} a_n$*. In case a sequence is not convergent, it is called divergent.*

Lemma 2.91 (Limits of Convergent Sequences) *Given sequences of real numbers* $(a_i)_{i\in\omega}$ *and* $(b_i)_{i\in\omega}$*, as well as a real number* $k \in \mathbb{R}$ *we have that*

$$k + \lim_{n\to\infty} a_n = \lim_{n\to\infty} (k + a_n) \tag{2.120}$$

$$k \cdot \lim_{n\to\infty} a_n = \lim_{n\to\infty} (k \cdot a_n) \tag{2.121}$$

$$\lim_{n\to\infty} a_n + \lim_{n\to\infty} b_n = \lim_{n\to\infty} (a_n + b_n) \tag{2.122}$$

$$\lim_{n\to\infty} a_n \cdot \lim_{n\to\infty} b_n = \lim_{n\to\infty} (a_n \cdot b_n) \tag{2.123}$$

Definition 2.92 (Monotonically Increasing Sequence) *A sequence of real numbers* $(a_n \in \mathbb{R})_{n\in\omega}$ *is monotonically increasing* **iff** $a_i \leqslant a_{i+1}$ *for all* $i \in \mathbb{N}_0$*.*

The real numbers $(\mathbb{R}, \leqslant)$ form a partial order; compare with Def 2.51. Obviously, a monotonically increasing sequence is an ω-chain in $(\mathbb{R}, \leqslant)$; compare with Def. 2.60. If a sequence of real numbers $(a_n \in \mathbb{R})_{n\in\omega}$ has an upper bound it is also said that it is bounded above. In real analysis, the least upper bound of a sequence of real numbers is usually called the supremum of the sequence; compare with Def. 2.53.

Theorem 2.93 (Monotone Convergence Theorem) *If a monotonically increasing sequence of real numbers* $(a_n \in \mathbb{R})_{n\in\omega}$ *is bounded above then its supremum exists and equals its limit, i.e.,*

$$\mathop{lim}_{n\to\infty}(a_n) = \sqcup\{a_n \mid n \in \mathbb{N}_0\} \tag{2.124}$$

2.5.5 Algebraic Structures

In Sect. 5.1.4 we will define operations $\oplus$ and $\otimes$ for functionals f of the form $f : S_n \to \ldots S_1 \to S_0 \to \mathbb{R}$ that mirror the arithmetic operations $+$ and $\times$ of the base domain $\mathbb{R}$ on the functional domain $S = S_n \to \ldots S_1 \to S_0 \to \mathbb{R}$. Then, it turns out that the algebraic structure $(S, \oplus, \otimes, 0_S)$ forms a vector space. This is important, because we need the entire set of vector space laws in proofs of properties of our denotational semantics. In this section we recap the vector space laws. As usual, we have organized the definition of vector spaces as a sequence of refinements, evolving from groups, over abelian groups and fields to vector spaces.

Definition 2.94 (Group) *An algebraic structure* $(G, +, 0_G)$ *with set* G*, operation* $+ : G \times G \to G$ *and element* 0_G*, called the neutral element, is called a group* **iff** *the following axioms hold true for all* $e, e', e'' \in G$*:*

$$(e + e') + e'' = e + (e' + e'') \tag{2.125}$$

$$e + 0_G = 0_G + e = e \tag{2.126}$$

$$\exists\, e^{-1} \in G \,.\, e + e^{-1} = e^{-1} + e = 0_G \tag{2.127}$$

The axiom in Eqn. (2.125) expresses the associativity of $+$. The axiom in Eqn. (2.126) expresses the neutrality of 0_G with respect to $+$ and axiom Eqn. (2.127) requires the existence of an inverse element e^{-1} for each element e in G.

Definition 2.95 (Abelian Group) *A group* $(G, +, 0_G)$ *is called an abelian group* **iff** *the following axiom of commutativity holds true for all* $e, e' \in G$*:*

$$e + e' = e' + e \tag{2.128}$$

Definition 2.96 (Field) *An algebraic structure* $(F, +, \cdot, 0_F, 1_F)$ *with set* F*, operations* $+ : F \times F \to F$ *and* $\cdot : F \times F \to F$*, and elements* $0_F \in F$ *and* $1_F \in F$ *is called a field* **iff** *both* $(F, +, 0_F)$ *and* $(F \backslash \{0_F\}, \cdot, 1_F)$ *are abelian groups plus the following axioms of distributivity hold true for all* $e, e', e'' \in F$*:*

$$e \cdot (e' + e'') = (e \cdot e') + (e \cdot e'') \tag{2.129}$$

$$(e + e') \cdot e'' = (e \cdot e'') + (e' \cdot e'') \tag{2.130}$$

Axioms Eqn. (2.129) and Eqn. (2.130) together express the distributiveness of $+$ over $\cdot$, where Eqn. (2.129) expresses left-distributiveness and Eqn. (2.130) expresses right-distributiveness. For us, the important example of a field is the field of real numbers $(\mathbb{R}, +, \cdot, 0, 1)$.

Definition 2.97 (Vector Space) *Given a field* $(F, +, \cdot, 0_F, 1_F)$ *based on a set* F*, called scalars. An algebraic structure* $(V, \oplus, \otimes, \mathbf{0}_V)$ *based on a set* V *of elements, called vectors, with operations* $\oplus : V \times V \to V$*, called vector addition, and* $\otimes : F \times V \to V$*, called scalar multiplication, as well as a neutral element* $\mathbf{0}_V \in V$*, called the null vector, is called a vector space over* F *or* F*-vector space for short* **iff** $(V, \oplus, \mathbf{0}_V)$ *forms an abelian group plus the following axioms hold true for all scalars* $i, i' \in F$ *and all vectors* $\mathbf{v}, \mathbf{v}' \in V$*:*

$$i \otimes (\mathbf{v} \oplus \mathbf{v}') = (i \otimes \mathbf{v}) \oplus (i \otimes \mathbf{v}') \tag{2.131}$$

$$(i + i') \otimes \mathbf{v} = (i \otimes \mathbf{v}) \oplus (i' \otimes \mathbf{v}) \tag{2.132}$$

$$(i \cdot i') \otimes \mathbf{v} = (i \otimes (i' \otimes \mathbf{v})) \tag{2.133}$$

$$1_F \otimes \mathbf{v} = \mathbf{v} \tag{2.134}$$

The laws expressed by Eqn. (2.131) and Eqn. (2.132) in Def. 2.97 are distributivity laws, i.e., Eqn. (2.131) explains how scalar multiplication distributes over vector addition, whereas Eqn. (2.132) explains how scalar addition distributes over vector multiplication. The axiom in Eqn. (2.133) expresses the compatibility of scalar multiplication with vector multiplication. Axiom Eqn. (2.134) expresses neutrality of the scalar one element 1_F with respect to scalar multiplication. Some further properties of vector spaces are given in the following Corollary 2.98.

Corollary 2.98 (Derived Vector Space Laws) *Given a vector space* $(V, \oplus, \otimes, \mathbf{0}_V)$ *over a field* $(F, +, \cdot, 0_F, 1_F)$ *the following holds for all scalars* $i \in F$ *and vectors* $\mathbf{v} \in V$*:*

$$0_F \otimes \mathbf{v} = \mathbf{0}_V \tag{2.135}$$

$$i \otimes \mathbf{0}_V = \mathbf{0}_V \tag{2.136}$$

The property expressed in Eqn. (2.135) explains the interplay of the scalar neutral element with vectors, whereas the property expressed in Eqn. (2.136) explains the interplay of the null vector with scalar values.

Often, we consider vector spaces $(V, \oplus, \otimes, \mathbf{0}_V)$ over a field $(F, +, \cdot, 0_F, 1_F)$ in which V is defined as a set of sequences $I \to F$ for some $I = \{1, \ldots, n\}$ or $I = \mathbb{N}$. These are the vectors in the narrow sense. Vectors of the form $(v_i)_{i \in \{1,\ldots,n\}}$ are called n-dimensional vectors, or just n-vectors for short. In case $V : \{1, \ldots, n\} \to F$ it is usual to consider V also as the n-times Cartesian product $F \times \ldots \times F$ and to denote it F^n as usual. We call n-vectors the finite vectors to distinguish them from the infinite vectors of the form $(v_i)_{i \in \mathbb{N}}$. In case of infinite vectors $V : \mathbb{N} \to F$ we denote V also as F^∞. We might also start counting n-vectors with zero, also less usual, so that an n-vector has the form $(v_i)_{i \in \{0,\ldots,n-1\}}$. Similarly, we also call sequences of the form $(v_i)_{i \in \omega}$ infinite vectors.

3

Syntax and Operational Semantics

In this chapter we introduce the probabilistic lambda calculus, which is a typed, probabilistic lambda calculus with recursion plus an extra construct for probabilistic choice. In this book we refer to this calculus simply as the probabilistic lambda calculus, i.e., the fact that the probabilistic lambda calculus is typed and has a recursion operator is taken for granted in this book. The probabilistic lambda calculus can be considered a minimalistic functional programming language. We completely specify the calculus, so that it could be immediately implemented as a working programming language. This means that we define both the syntax and the operational semantics of the calculus. The operational semantics is given in two stages. In the first stage, we give a labeled transition system between terms of the language. On the basis of this we elaborate a Markov chain semantics of the calculus. For this purpose, we will use the labeled transition system as a probabilistic matrix and will then explain reduction probabilities as hitting probabilities in the Markov chain that exists for this probabilistic matrix. Note that already the first stage of this semantics, i.e., the labeled transition system, is sufficient as a complete specification for the implementation of a programming language. However, we are interested in properties of programs. The purpose of introducing a formal language is to enable formal reasoning about program properties. Now, it is the second stage that makes the difference between simulation and analytical treatment of program properties. Only with the full Markov chain semantics we gain access to the important mathematical toolkit that we desire for semantic analysis. We will encounter this, e.g., when we investigate the termination behavior of the probabilistic lambda calculus in Chap. 4.

We introduce the probabilistic lambda calculus as an extension of the standard typed lambda calculus. The standard operational semantics of the typed lambda calculus is given by a transition system. This transition system is then extended to a labeled transition system. Here the labels carry the probabilities of probabilistic choices.

3.1 Syntax of the Probabilistic Lambda Calculus

3.1.1 Context-Free Syntax

We define the set of types T by the following grammar:

$$t = num \quad | \quad bool \quad | \quad t \to t$$

We use T_g to denote the set of ground types num and $bool$. Types of the form $t_1 \to t_2$ are called higher types.

We assume a countable set Var containing arbitrarily many typed variable symbols for each type $t \in T$. As usual, we do not fix a concrete syntax for variable symbols but simply assume that we have enough of them, e.g., x_t, y_t, z_t or $v_t, v_t'', v_t''', \ldots$ for each type $t \in T$. Furthermore, as usual, we feel free to omit the type subscript of variable symbols wherever the type can be implicitly derived from the typing rules in Sect. 3.1.2. We define a set C_{num} of constant symbols n_i for each natural number $i \in \mathbb{N}_0$. In example terms, we feel free to use the usual number symbols as these constants. We define a set C_{bool} of constant symbols $\dot{t}$ and $\dot{f}$ to denote the corresponding truth constants. We define the set of constants C as the union of C_{num} and C_{bool}.

We define the set of lambda terms Λ according to the following grammar:

$$M = x_t \in Var \mid \tag{3.1}$$

$$n_i \in C_{num} \mid +1(M) \mid -1(M) \mid \tag{3.2}$$

$$\dot{t} \mid \dot{f} \mid 0?(M) \mid \tag{3.3}$$

$$if\, M\; then\; M\; else\; M \mid \tag{3.4}$$

$$\lambda x_t.M \mid MM \mid \mu M \tag{3.5}$$

$$M = M|M \tag{3.6}$$

Terms of the form $+1(M)$ and $-1(M)$ stand for the addition resp. substraction of one, whereas $0?(M)$ is a test for zero. Terms of the form $if M then N_1 else N_2$ are called conditional expressions or conditionals for short. Throughout the book we also use $if(M, N_1, N_2)$ as an alternative concrete syntax for conditionals. The term M in a conditional expression is called its condition, and the terms N_1 and N_2 are called its clauses, i.e., its left resp. right clause. Terms of the form MN are called applications. The term M in an application is called the function and the term N is called the parameter of the application. Terms of the form $\lambda v.M$ are called lambda abstractions or abstractions for short. An abstraction always has a higher type, i.e., it always defines a function. Whenever we refer to an abstraction $\lambda v.M$ as a function, it is usual to call the term M its function body or body for short. Terms of the form μM are called μ-recursions or recursions, where each μ is called the recursion operator, or μ-operator. Furthermore, the μ-operator is sometimes also called the fixed-point operator. However, we prefer to use the term fixed-point operator rather for the denotational counter part Φ of the μ-operator; see Def. 2.80.

Last but not least, terms of the form $M|N$ are called probabilistic choices and the corresponding operator $|$ is called the choice operator. Please, just as a caution, note that we use the vertical bar $|$ in Eqns. (3.1) through (3.5) as the usual separator for the several grammatical rules. However, in Eqn. (3.6) the vertical bar in $M|M$ is our syntax for the choice operator and not a rule separator. The choice to use $M|N$ to denote probabilistic choice is an arbitrary choice. Other options that might be be found in the literature are, e.g., $M \oplus N$ or the fat bar notation $M\|N$.

The probabilistic lambda calculus Λ is the language specified by the whole set of rules in Eqns. (3.1) through (3.6). The standard typed lambda calculus, let us denote it by Λ', is the subset of Λ which is specified by the Eqns. (3.1) through (3.5), i.e., the probabilistic lambda calculus is distinguished from the standard calculus by exactly one extra language construct, the probabilistic choice, introduced by the rule in Eqn. (3.6). In the sequel, we assume that all syntactical definitions that we will give as well as all the syntactical properties that we discuss for the probabilistic lambda calculus Λ such as free and bound variables, open and closed terms, ground terms etc. immediately transport to the standard typed lambda calculus Λ'.

As usual, the grammar given by Eqn. (3.1) *ff.* is considered an abstract syntax specification. This means that the objects in Λ are actually abstract syntax trees. As usual, we do not stress that fact any more in the sequel and simply call the objects in Λ terms. Nevertheless, implicitly we exploit the fact that our terms are abstract syntax trees heavily in structural inductive proofs throughout the book. As usual, we use parentheses in concrete term syntax to indicate the term tree syntax of the abstract syntax. Also, we are allowed to omit parentheses and can then assume the usual constructor precedences and fixities. Application binds tighter then λ- and μ-abstraction. Application also binds tighter then the case construct in its third argument, i.e., we have $\mathit{if}\, T\ \mathit{then}\ N_1\ \mathit{else}\ MN$ equals $\mathit{if}\, T\ \mathit{then}\ N_1\ \mathit{else}\ (MN)$. Application associates to the left, i.e., $M_0 M_1 M_2$ should be parsed as $(M_0 M_1) M_2$. Consequently, the type constructor associates to the right, i.e., $t_1 \to t_2 \to t_3$ should be parsed as $t_1 \to (t_2 \to t_3)$.

Next, we need to define precedence and associativity for the choice operator. The choice constructor binds even tighter than application, and therefore also binds tighter than λ- and μ-abstraction as well as conditional expressions. For example, $\lambda x_s.M_1 M_2|M_3$ should be parsed as $\lambda x_s.(M_1(M_2|M_3))$. We define, as an arbitrary choice, that the choice constructor associates to the left, i.e., $L|M|N$ should be parsed as $(L|M)|N$. In Corollary 3.8 we will see that the semantics of the choice constructor is non-associative, and therefore fixing a precedence rule for the choice constructor matters.

3.1.2 Typing Rules

A lambda term M is well typed and then has type t, denoted by $M : t$, according to the following definition of type derivation:

$$\frac{x_t \in Var}{x_t : t} \tag{3.7}$$

$$\frac{n_i \in C_{num}}{n_i : num} \quad \frac{M : num}{+1(M) : num} \quad \frac{M : num}{-1(M) : num} \tag{3.8}$$

$$\frac{}{\dot{t} : bool} \quad \frac{}{\dot{f} : bool} \quad \frac{M : num}{0?(M) : bool} \tag{3.9}$$

$$\frac{M : bool \quad N_1 : t \quad N_2 : t}{\mathit{if}\, M \;\mathit{then}\; N_1 \;\mathit{else}\; N_2 \;:\; t} \tag{3.10}$$

$$\frac{x_{t_1} \in Var \quad M : t_2}{\lambda x_{t_1}.M : t_1 \to t_2} \quad \frac{M : t_1 \to t_2 \quad N : t_1}{MN : t_2} \tag{3.11}$$

$$\frac{M : t \to t}{\mu M : t} \tag{3.12}$$

$$\frac{M : t \quad N : t}{M|N : t} \tag{3.13}$$

It is usual to define typing of terms as a type derivation system as given by Eqn. (3.7) *ff*; compare, e.g., with [223, 19, 118, 42]. Formally, the so-called typing rules in Eqn. (3.7) *ff.* are rule sets that inductively define the type relation $: \subseteq \Lambda \times T$; compare with our discussion of inductive definitions in Sect. 2.4 and, in particular, to the discussion of rule-based inductive definitions in Sect. 2.4.2. Now, a term M is considered well typed if there exists a type $t \in T$ such that $M : t$. Then, M is said to have type t. Actually, it can be shown that the type of a term, if it exists, is unique, i.e., that the type relation is a partial function.

As expressed by the next lemma, each type has a right-associate top-level structure in terms of a sequence of other types, finally targeting a basic type. This canonical structure of types can be exploited in proofs of properties of terms or types. Here, it allows for a natural induction over the length of this canonical type structure, which is an alternative to the usual structural induction over the construction of terms.

Lemma 3.1 (Top-Level Functional Structure of Types) *Each type $t \in T$ has the form $(t_n \to (\ldots(t_2 \to (t_1 \to g))...))$ for an $n \geqslant 0$, a type $t_i \in T$ for each $1 \leqslant i \leqslant n$ and a ground type $g \in T_g$.*

Proof. By structural induction over the set of terms T. □

In case $n = 0$ in Lemma 3.1 the type t takes the form $g \in T_g$. For $n = 1$ it takes the form $t \to g$ with $t \in T$ and $g \in T_g$.

Typing Rules and Type Environments

Our typing rules are straightforward; for example, they directly correspond to the typing rules given in [223]. We just use some other notation than [223] to denote the rules, which arranges the premise and the conclusion vertically, separated by a horizontal line. Such notation is common for rule-based inductive definitions, compare with Sect. 2.4.2, and, actually, it also very common in the type systems community; compare with [210, 42, 219]. In particular, we work without so-called type environments. In many texts [118, 42] type environments are used to give types to variables. In such formalizations, a type statement has the form $\Gamma \vdash M : t$ for some type environment Γ, some term M and some type t, whereas our type statements simply have the form $M : t$. In formalizations without type environments we work with a set of typed variables, which shows in the type subscript notation x_t in our calculus. In a formalization with type environments, we start with a single set of untyped variables that receive their types from the type environment. Therefore, a type environment Γ is a set of basic typing tuples of the form $x : t$. For example, compare the very basic variable-typing rule Eqn. (3.7) with its counterpart version based on a type environment, which would look like this:

$$\frac{}{\{x : t\} \vdash x : t} \tag{3.14}$$

Type environments are particularly useful for formalizing more advanced type system concepts [42, 219] such as polymorphism, type genericity or existentially quantified types. Also we use type environment notation on other occasions, e.g., when we specify type safety for the generative programming language GENOUPE [178, 90, 91, 92] or when we specify the strongly typed web service programming language NSP [81, 83]. On other occasions we also deal without type environments, e.g., in case of the type systems for the web service type recovery tool JSPICK [87] or the reflective constraint language OCL_R [79]. In the current book type environments offer no advantage and therefore we stay with a formalization that works without type environments.

3.1.3 Closed Terms, Open Terms, Programs and Values

The notions of variables of a term, free and bound variables, closed and open terms as well as programs are all standard and defined as usual. Given a term M, we denote the set of its variables by $Var(M)$, the set of its free variables by $V_{free}(M)$ and the set of its bound variables as $V_{bound}(M)$. We denote the set of all closed terms by $\Lambda_\emptyset$ and the set of all programs by Λ_P. It is the task of this section to give definitions of all of these concepts, plus some further important syntactical concepts such as variable substitution, α-equivalence and values.

First, in order to ease the definition of the several syntactical constructs we introduce a set Ψ of functions on Λ of different arity that we call syntactical constructors or also just constructors for short.

Definition 3.2 (Syntactical Constructors) *The set Ψ of syntactical constructors consists of the following functions:*

$$
\begin{aligned}
\psi_+ : \Lambda \rightarrow \Lambda &=_{DEF} M \mapsto +1(M) \\
\psi_- : \Lambda \rightarrow \Lambda &=_{DEF} M \mapsto -1(M) \\
\psi_? : \Lambda \rightarrow \Lambda &=_{DEF} M \mapsto 0?(M) \\
\psi_{if} : \Lambda \times \Lambda \times \Lambda \rightarrow \Lambda &=_{DEF} \langle L, M, N \rangle \mapsto \mathit{if}\, L \,\mathit{then}\, M \,\mathit{else}\, N \\
\psi_{app} : \Lambda \times \Lambda \rightarrow \Lambda &=_{DEF} \langle M, N \rangle \mapsto MN \\
\psi_\lambda : \Lambda \times \Lambda \rightarrow \Lambda &=_{DEF} \langle v, M \rangle \mapsto \lambda v.M \\
\psi_\mu : \Lambda \rightarrow \Lambda &=_{DEF} M \mapsto \mu M \\
\psi_\gamma : \Lambda \times \Lambda \rightarrow \Lambda &=_{DEF} \langle M, N \rangle \mapsto M|N
\end{aligned}
\tag{3.15}
$$

The constructors Ψ can be exploited to avoid redundancies in the definition of syntactical properties. Furthermore, we can use constructor notation also to shorten proofs, because they allow us to apply the same argument simultaneously to different kinds of terms at once; see, e.g., the proof of the environments-shortening Lemma, i.e., Lemma 5.23. The constructors decompose the set of lambda terms Λ, to be more precise, the set Λ excluding variables and constants. This means that the images of the constructors are pairwise disjoint and, furthermore, the union of the images of all constructors includes all terms except variables and constants. This follows from the grammar of Λ as given by the rules in Eqn. (3.1) *ff.* and the definition of Ψ in Def. 3.2. Each rule in the grammar except those for variables and constants is uniquely represented by one syntactical constructor in Ψ.

Let us proceed with the definition of the needed syntactical concepts. We start with the concept of variables $\mathit{Var}(M)$ of a term M. In case of variables $v \in \mathit{Var}$ and constants $c \in C$ the set of variables of a term is defined as follows:

$$\mathit{Var}(v) =_{DEF} \{v\} \tag{3.16}$$

$$\mathit{Var}(c) =_{DEF} \emptyset \tag{3.17}$$

For all other terms, i.e., terms M such that $M = \psi(M_1, \ldots, M_{n_\psi})$ for some constructor $\psi \in \Psi$ and appropriately many terms $M_1, \ldots M_{n_\psi}$ the set of variables is defined as follows:

$$\mathit{Var}(\psi(M_1, \ldots, M_{n_\psi})) =_{DEF} \mathit{Var}(M_1) \cup \ldots \cup \mathit{Var}(M_{n_\psi}) \tag{3.18}$$

Let us turn to the concepts of free variables and bound variables of a given term. In case of variables $v \in \mathit{Var}$ and constants $c \in C$ the set of free variables and the set of bound variables of a term are defined as follows:

$$V_{\mathit{free}}(v) =_{DEF} \{v\} \tag{3.19}$$

$$V_{\mathit{bound}}(v) =_{DEF} \emptyset \tag{3.20}$$

$$V_{\mathit{free}}(c) =_{DEF} \emptyset \tag{3.21}$$

$$V_{\mathit{bound}}(c) =_{DEF} \emptyset \tag{3.22}$$

For lambda abstractions $\lambda v.M$ the sets of its free variables and its bound variables are defined as follows:

$$V_{\mathit{free}}(\lambda v.M) \equiv_{DEF} V_{\mathit{free}}(M)\backslash\{v\} \tag{3.23}$$

$$V_{bound}(\lambda v.M) \equiv_{DEF} V_{bound}(M) \cup \{v\} \tag{3.24}$$

For all terms other than variables, constants or lambda abstractions, i.e., terms M such that M equals $\psi(M_1,\ldots,M_{n_\psi})$ for some constructor $\psi \neq \psi_\lambda$ and appropriately many terms $M_1,\ldots M_{n_\psi}$ the sets of their free variables and their bound variables are defined as follows:

$$V_{\mathit{free}}\big(\psi(M_1,\ldots,M_{n_\psi})\big) =_{DEF} V_{\mathit{free}}(M_1) \cup \ldots \cup V_{\mathit{free}}(M_{n_\psi}) \tag{3.25}$$

$$V_{bound}\big(\psi(M_1,\ldots,M_{n_\psi})\big) =_{DEF} V_{bound}(M_1) \cup \ldots \cup V_{bound}(M_{n_\psi}) \tag{3.26}$$

Now, we are able to define the notions of closed and open terms. A term $M \in \Lambda$ is a closed term if and only if it contains no free variables, i.e., $V_{\mathit{free}}(M) = \emptyset$. Otherwise, it is called an open term. Similarly, the set of closed terms $\Lambda_\emptyset$ and the set of open terms Λ_{open} are defined as follows:

$$\Lambda_\emptyset =_{DEF} \{M \in \Lambda \mid V_{\mathit{free}}(M) = \emptyset\} \tag{3.27}$$

$$\Lambda_{open} =_{DEF} \{M \in \Lambda \mid V_{\mathit{free}}(M) \neq \emptyset\} \tag{3.28}$$

Next, we define the set of ground terms, denoted by Λ_g, and the set of higher-typed terms, denoted by $\Lambda_\rightarrow$, as follows:

$$\Lambda_g =_{DEF} \{M \in \Lambda \mid \exists t \in T_g \,.\, M : t\} \tag{3.29}$$

$$\Lambda_\rightarrow =_{DEF} \{M \in \Lambda \mid \exists t, t' \in T \,.\, M : t \rightarrow t'\} \tag{3.30}$$

Next, a program is a closed term of ground type. We denote the set of programs by Λ_P, i.e.,

$$\Lambda_P =_{DEF} \Lambda_\emptyset \cap \Lambda_g \tag{3.31}$$

Variable Substitution

Next, we turn to the important notion of variable substitution. Substitution is the essential notion of term rewriting, term rewriting systems and symbolic computation. Given terms M and N, substitution is about replacing all occurrences of a free variable v in term M by the new expression N. We denote such substitution by the common notation $M[v := N]$. Given $M[v := N]$ we also say that v is substituted by N. Different forms of variable substitution exist. What we actually introduce here is a so-called context substitution, which means that free variables in N might be captured by λ-bindings and this way be turned into bound variables after substitution. Context substitution is a straightforward notion of substitution. Several refined notions of variable substitution exist that circumvent the variable capture problem. In this book, we will always only substitute variables with closed terms, so that we do not run the risk that free variables become captured by substitution; see also Sect. 3.2.1, where we explain this in more depth. For this reason, we just stay with context substitution. Also, we just use context substitution, variable substitution and substitution as synonyms in this book.

For variables $v, v' \in \mathit{Var}$, constants $c \in C$ and closed terms $N \in \Lambda_\emptyset$ substitution is defined as follows:

$$v[v{:=}N] \equiv_{DEF} N \tag{3.32}$$
$$v[v'{:=}N] \equiv_{DEF} v \quad , v \neq v' \tag{3.33}$$
$$c[v{:=}N] \equiv_{DEF} c \tag{3.34}$$

For lambda abstractions of the form $\lambda v.M \in \Lambda$, variables $v, v' \in \mathit{Var}$ and terms $N \in \Lambda$ substitution is defined as follows:

$$\lambda v.M[v{:=}N] \equiv_{DEF} \lambda v.M \tag{3.35}$$
$$\lambda v.M[v'{:=}N] \equiv_{DEF} \lambda v.(M[v'{:=}N]) \quad , v \neq v' \tag{3.36}$$

For all other terms, i.e., terms M such that $M = \psi(M_1, \ldots, M_{n_\psi})$ for some constructor $\psi \neq \psi_\lambda$ and appropriately many terms $M_1, \ldots M_{n_\psi}$ substitution is defined as follows:

$$\psi(M_1, \ldots, M_{n_\psi})[v{:=}N] =_{DEF} \psi(M_1[v{:=}N], \ldots, M_{n_\psi}[v{:=}N]) \tag{3.37}$$

α-Equivalence

For the sake of completeness, we also introduce the notion of α-equivalence. An understanding of this notion is necessary to follow the discussion on substitution and the capturing of free variables that we conduct in Sect. 3.2.1. The α-equivalence relation is intended to compare lambda terms that are syntactically equal up to renaming of bound variables. It is defined as the least relation $\equiv_\alpha$ for which the following holds true for all variables $v \in \mathit{Var}$, constants $c \in C$, lambda abstractions $\lambda x.M$, $\lambda y.N$, and terms of the form $\psi(M_1, \ldots, M_{n_\psi})$, $\psi(N_1, \ldots, N_{n_\psi})$ for some $\psi \neq \psi_\lambda$ and appropriately many terms $M_1, \ldots, M_{n_\psi}$ and $N_1, \ldots, N_{n_\psi}$:

$$v \equiv_\alpha v \qquad c \equiv_\alpha c \tag{3.38}$$

$$\lambda x.M \equiv_\alpha \lambda y.N \ \textbf{iff} \ \exists z \notin \mathit{Var}(M \cup N) \,.\, M[x := z] \equiv_\alpha N[y := z] \tag{3.39}$$

$$\psi(M_1, \ldots, M_{n_\psi}) \equiv_\alpha \psi(N_1, \ldots, N_{n_\psi}) \ \textbf{iff} \ \forall 1 \leqslant i \leqslant n_\psi \,.\, M_i \equiv_\alpha N_i \tag{3.40}$$

Values of the Probabilistic Lambda Calculus

Next, we turn to the concept of values. The values of the probabilistic lambda calculus, denoted by Λ_V, are defined as the set consisting of constants and closed lambda abstractions, i.e.,

$$\Lambda_V = C \,\cup\, \{\, \lambda v.B \mid B \in \Lambda, V_{free}(B) = \{v\}\,\} \tag{3.41}$$

It is immediately clear to name the constants of a programming calculus as values, because we consider the constants as the result of reductions. A program run is considered to terminate if it eventually reaches a constant. However, a broadened notion of values is often needed in the definition of concrete evaluation strategies for a programming calculus or programming language. For example, in a call-by-value operational semantics, we will defer a function application until the argument is reduced to a given notion of value. Requiring that these values are constants could then exclude some meaningful programs from terminating. The values of the call-by-value evaluation strategy for the standard typed lambda calculus are exactly as defined here, i.e., constants plus lambda abstractions. In this book we deal with call-by-name operational semantics only; please also compare with the discussion on that in Sect. 3.2. Therefore, we do not need a notion of values in the definition of the operational semantics. Nevertheless, the notion of values as given in Eqn. (3.41) is crucial for us. We need it in the analysis of denotational semantics and its correspondence to operational semantics in Sect. 5.3. More concretely, we need values to define a big-step evaluation semantics for our calculus; compare with Lemma 5.30.

3.2 Operational Semantics of the Typed λ-Calculus

Before we turn to the operational semantics of the probabilistic lambda calculus Λ in Sect. 3.3 we first digress with an investigation of the standard typed lambda calculus Λ'. We give an operational semantics in the standard way; see, e.g., [223], as a so-called reduction relation between the closed terms of Λ'. We do so, as usual, by specifying an immediate reduction relation that is then turned into the desired reduction relation. For the proceedings of the book it is helpful to have the operational semantics of the standard typed lambda calculus to hand, for instructive purposes and, even more important, for better comparability. The section can be skipped on first reading without loss, if the reader is already familiar with the typed lambda calculus and its operational semantics.

The so-called immediate reduction relation is introduced as a relation between the closed lambda terms, i.e.,

$$\rightarrow \subseteq \Lambda'_\emptyset \times \Lambda'_\emptyset \tag{3.42}$$

The immediate reduction relation is defined inductively by the following sets of rules:

$$i)\frac{M \to M'}{+1(M) \to +1(M')} \qquad ii)\frac{i \geqslant 0}{+1(n_i) \to n_{(i+1)}} \tag{3.43}$$

$$i)\frac{M \to M'}{-1(M) \to -1(M')} \qquad ii)\frac{i \geqslant 1}{-1(n_i) \to n_{(i-1)}} \qquad iii)\frac{}{-1(n_0) \to n_0} \tag{3.44}$$

$$i)\frac{M \to M'}{0?(M) \to 0?(M')} \qquad ii)\frac{}{0?(n_0) \to \dot{t}} \qquad iii)\frac{i \geqslant 1}{0?(n_i) \to \dot{f}} \tag{3.45}$$

$$\frac{M \to M'}{\mathit{if}(M, N', N'') \to \mathit{if}(M', N', N'')} \tag{3.46}$$

$$i)\frac{}{\mathit{if}(\dot{t}, N', N'') \to N'} \qquad ii)\frac{}{\mathit{if}(\dot{f}, N', N'') \to N''} \tag{3.47}$$

$$i)\frac{M \to M'}{MN \to M'N} \qquad ii)\frac{}{(\lambda v.B)N \to B[v := N]} \tag{3.48}$$

$$\frac{}{\mu M \to M(\mu M)} \tag{3.49}$$

Now, the reduction relation $\overset{\star}{\to}$ is defined as the reflexive and transitive closure of the immediate reduction relation $\to$. It turns out that reduction to constants is uniquely given. This means that given a program $M \in \Lambda'$ and a constant $c \in C$ such that $M \overset{\star}{\to} c$ we have that c is uniquely determined, i.e., there is no other constant $c' \in C$ with $c' \neq c$ and $M \overset{\star}{\to} c'$. Based on that, the evaluation semantics of the standard typed lambda calculus can be defined as a partial function that assigns to each program $M \in \Lambda'_P$ the uniquely given constant c to which it can be reduced via $M \overset{\star}{\to} c$, given that this c exists at all. The evaluation semantics necessarily is a partial function, because we also have non-terminating programs in Λ'_P. A non-terminating program in Λ'_P is a program that never reaches a constant.

Call-by-Name Operational Semantics

The operational semantics as specified by the rules (3.43) *ff.* is a call-by-name semantics. This shows in the reduction rule Eqn. (3.48-ii). If a reduction has reached an intermediate term $(\lambda v.B)N$ the term N is immediately, without further reduction, substituted as an actual parameter into the function body B. This is exactly what call-by-name is about. Other evaluation strategies are possible [220, 69]. With call-by-value semantics substitution $(\lambda v.B)N$ of the argument is deferred until the argument has been reduced further to some value, i.e., $(\lambda v.B)N \overset{\star}{\to} (\lambda v.B)N' \to B[v := N']$ for some value $N' \in \Lambda'_V$, where Λ'_V stands, as usual, for the set consisting of all constants C plus all lambda abstractions.

3.2.1 The Variable Capture Problem

The definition of the operational semantics by the rules (3.43) *ff.* is based on a concept of variable substitution $M[v := N]$; see the rule set (3.48-ii). See how we have defined this syntactical concept in Eqn. (3.32) *ff.* The substitution we use is a very general form of substitution, i.e., a so-called context substitution. This means that free variables of an argument term may become captured underneath a lambda binding after substitution, which would change the intended semantics of the term. The point is that this fact will never pose a problem in our operational semantics. In the operational semantics of the typed lambda calculus, the only potentially problematic rule set is the rule set (3.48-ii) which reveals a redex of the form $(\lambda v.B)N$. However, the argument N cannot have a free variable, because the reduction relation is defined for closed terms only. Now, given that the whole application term $(\lambda v.B)N$ is closed, we also know that the argument term N is closed; compare with the definition of free variables in Eqn. (3.20) *ff.*

In general, reductions for λ-calculi are defined for all terms, not only for closed terms. Furthermore, it is usual to consider transition relations that are compatible closures. A motivation for using a transition relation that is a compatible closure is to avoid prescribing reduction strategies from the outset and therefore enable the study of different reduction strategies. For an explanation of compatible closure, see also [18], §3.1.1. Compatible closure means that a notion of reduction such as the β-reduction $(\lambda v.M) \to M[x := N]$ is transferred to arbitrary sub-terms of any other term, i.e., that given terms M and N we have that $M \to N$ implies $C[M] \to C[N]$. Here, $C[M]$ is a usual notation that stands for a context substitution without explicitly mentioning a variable as in the notation $C[v := M]$. Rather, the term $C[\_]$ is called a context with a so-called hole $[\_]$. Let us have a look at an example in which a variable capture problem might occur if a plain context substitution is used without further arrangements. Let us consider the term $(\lambda x.(\lambda y.\lambda x.y)x)\ 0\ 1$ Now, with the reduction system as established with call-by-name operational semantics, it is possible to reduce this term as follows:

$$\begin{aligned} &(\lambda x.(\lambda y.\lambda x.y)x)\ 0\ 1 \\ &\to (\lambda y.\lambda x.y)\ 0\ 1 \\ &\to (\lambda x.0)1 \\ &\to 0 \end{aligned} \tag{3.50}$$

Now, if we allow arbitrary β-reduction on arbitrary sub-terms including open terms based on context substitution, the following other reduction also becomes possible:

$$\begin{aligned} &(\lambda x.(\lambda y.\lambda x.y)x)\ 0\ 1 \\ &\to (\lambda x.\lambda x.x)\ 0\ 1 \\ &\to (\lambda x.x)1 \\ &\to 1 \end{aligned} \tag{3.51}$$

The problem shows in the fact that the two computations in Eqns. (3.50) and (3.51) yield different outcomes, i.e., 0 and 1. Here, the outcome 1 must be considered spurious, which shows in the unintended reduction of the subterm $(\lambda y.\lambda x.y)x$ to the term $\lambda x.x$. There exist different approaches to deal with the variable capture problem. A typical approach is to introduce an improved concept of variable substitution that avoids the capture of variables from the outset. This can be achieved by adding an appropriate side condition $V_{free}(N) \cap V_{bound}(M) = \emptyset$ to the relevant clause in the definition of substitution, compare with Eqn. (3.36), and then allow for renaming of variables, i.e., the introduction of fresh variables during substitution. Technically, this can be accompanied by a factorization of the set of terms along α-equivalence as, e.g., in [118]. Another technical option is to establish a variable-name-free apparatus like the de Bruijn notation [65].

In this book we are not concerned with the variable capture problem and therefore stay with context substitution. This eases argumentations and formal proofs, in which we do not have to care about extra syntactical properties and technical machinery for variable substitution.

3.3 The Probabilistic Operational Semantics

3.3.1 The One-Step Semantics

We introduce the call-by-name one-step semantics to define the probabilities with which immediate transitions between closed terms are possible. We use $M \xrightarrow{i} N$ to denote that a transition from M to N is possible with probability i. In that sense, the one-step semantics can be considered as a labeled transition system. The call-by-name one-step semantics is introduced as a relation:

$$\rightarrow \subseteq \Lambda_\emptyset \times \Lambda_\emptyset \times \{0, 0.5, 1\} \tag{3.52}$$

Then, we introduce $M \xrightarrow{i} N$ as a notation for a predicate on triples of the one-step semantics in the following way:

$$M \xrightarrow{i} N \equiv_{DEF} (\langle M, N, i \rangle \in \rightarrow) \tag{3.53}$$

In due course, with Lemma 3.3, it will turn out that the one-step semantics is a total function, i.e., $\rightarrow\colon \big(\Lambda_\emptyset \times \Lambda_\emptyset \longrightarrow \{0, 0.5, 1\}\big)$. For the time being, we treat the one-step semantics as a relation as defined in Eqn. (3.52). Now, we define the one-step semantics inductively by the rule sets given in Eqns. (3.54) through (3.64) for all closed lambda terms M, M', N_1, N_2, N'. We have grouped the rule sets into several blocks. The first two blocks deal with term constructs that make up usual typed lambda calculi with recursion. We start with a block of rule sets needed for basic operators and conditional expressions:

$$i)\frac{M \notin C \quad M \xrightarrow{i} M'}{+1(M) \xrightarrow{i} +1(M')} \qquad ii)\frac{i \geqslant 0}{+1(n_i) \xrightarrow{1} n_{(i+1)}} \tag{3.54}$$

$$i)\frac{M \notin C \quad M \xrightarrow{i} M'}{-1(M) \xrightarrow{i} -1(M')} \qquad ii)\frac{i \geqslant 1}{-1(n_i) \xrightarrow{1} n_{(i-1)}} \qquad iii)\frac{}{-1(n_0) \xrightarrow{1} n_0} \tag{3.55}$$

$$i)\frac{M \notin C \quad M \xrightarrow{i} M'}{0?(M) \xrightarrow{i} 0?(M')} \qquad ii)\frac{}{0?(n_0) \xrightarrow{1} \dot{t}} \qquad iii)\frac{i \geqslant 1}{0?(n_i) \xrightarrow{1} \dot{f}} \tag{3.56}$$

$$\frac{M \notin C \quad M \xrightarrow{i} M'}{\mathit{if}(M, N', N'') \xrightarrow{i} \mathit{if}(M', N', N'')} \tag{3.57}$$

$$i)\frac{}{\mathit{if}(\dot{t}, N', N'') \xrightarrow{1} N'} \qquad ii)\frac{}{\mathit{if}(\dot{f}, N', N'') \xrightarrow{1} N''} \tag{3.58}$$

The next three rules deal with the semantics of function application and recursion:

$$i)\frac{M \neq (\lambda v.B) \quad M \xrightarrow{i} M'}{MN \xrightarrow{i} M'N} \qquad ii)\frac{}{(\lambda v.B)N \xrightarrow{1} B[v := N]} \tag{3.59}$$

$$\frac{}{\mu M \xrightarrow{1} M(\mu M)} \tag{3.60}$$

The third block of derivation rules deals with the semantics of the choice operator that is the key ingredient of the probabilistic lambda calculus:

$$i)\frac{M \neq N}{M|N \xrightarrow{0.5} M} \qquad ii)\frac{M \neq N}{M|N \xrightarrow{0.5} N} \qquad iii)\frac{}{M|M \xrightarrow{1} M} \tag{3.61}$$

The next two blocks of derivation rules are needed to make the one-step semantics a total function. Actually, the derivation rules ensure that the one-step semantics forms a probabilistic matrix. This way, the one-step semantics can serve as the basis of a Markov chain semantics. The following two rules deal with transition probabilities for the values:

$$i)\frac{M \in C}{M \xrightarrow{1} M} \qquad ii)\frac{}{\lambda v.B \xrightarrow{1} \lambda v.B} \tag{3.62}$$

The next two rules introduce the necessary null transition probabilities for all relevant terms:

$$\frac{M \xrightarrow{1} N \quad N' \neq N}{M \xrightarrow{0} N'} \tag{3.63}$$

$$\frac{M \xrightarrow{0.5} N_1 \quad M \xrightarrow{0.5} N_2 \quad N' \neq N_1 \quad N' \neq N_2}{M \xrightarrow{0} N'} \tag{3.64}$$

With rule (3.62-i) we ensure a direct transition $M \xrightarrow{1} M$ between each constant M and itself. This makes sense. It means that we will model a terminating computation run as an infinite *process instance* that is *absorbed* behind the *hit* of a constant. This is a perfectly adequate model of the scenario, because from an observer's viewpoint infinite process instances that are absorbed behind constants can be considered conceptually or abstractly finite, in particular, because the hit of a constant can be run-time decided by each computational engine that is used to run our programs.

For any two terms M and N there exists exactly one number i such that $M \xrightarrow{i} N$ belongs to the one-step semantics. In the standard typed lambda calculus, impossibility of a reduction between two terms M and N simply shows in the fact, that there is no tuple $M \rightarrow N$ in the immediate reduction relation; compare with Sect. 3.2. In the probabilistic lambda calculus, the impossibility of a reduction is turned into a zero percent probability.

Now, the extra one-step reductions introduced by (3.62) through (3.64) come at a price, i.e., that we need to introduce further side conditions into the premises of rule sets as compared to the standard typed lambda calculus. Compare, as an example, rule (3.54-i) and its corresponding rule (3.43-i). In rule (3.54-i) we must add the extra side condition $M \notin C$. Without this extra side condition we would be able to derive $+1(n_0) \xrightarrow{1} +1(n_0)$ via rules (3.62-i) and (3.54-i), which is unwanted as an immediate reduction. Also note that we already have $+1(n_0) \xrightarrow{1} n_1$ via rule (3.54-ii) and therefore $+1(n_0) \xrightarrow{0} +1(n_0)$ via (3.63), which is completely adequate. To see this, you might also want to have a look at Lemma 5.48, which is a refinement of Lemma 3.3 and which, amongst other things, characterizes the one step-semantics as follows: for each closed term P there either exists a closed term P' so that $P \xrightarrow{1} P'$ and $P \xrightarrow{0} P''$ for all $P'' \neq P'$ or otherwise there exist closed terms P' and P'' so that $P' \neq P''$, $P \xrightarrow{0.5} P'$ and $P \xrightarrow{0.5} P''$ and $P \xrightarrow{0} P'''$ for all P''' with $P''' \neq P'$ and $P''' \neq P''$.

Lemma 3.3 (The One-Step Semantics is a Function) *The one-step semantics is a total function on pairs of closed terms* $\rightarrow : (\Lambda_\emptyset \times \Lambda_\emptyset \longrightarrow \{0, 0.5, 1\})$ *that fulfills the equation Eqn. (3.65) for all* $X, X' \in \Lambda_\emptyset$*, where* $X \rightarrow X'$ *is used to denote* $\rightarrow (X, X')$*.*

Proof. We need to show that the relation $\rightarrow : \Lambda_\emptyset \times \Lambda_\emptyset \times \{0, 0.5, 1\}$ is right-unique and left-total in $\Lambda_\emptyset \times \Lambda_\emptyset$. This can be shown by a complete case distinction for all pairs of closed terms $\langle X, X' \rangle$ in $\Lambda_\emptyset \times \Lambda_\emptyset$. It is possible to structure the complete case distinction along the possible cases of syntax construction of the first function argument X of $\langle X, X' \rangle$. We only show one case that already suffices to demonstrate how this works, i.e., the case of constants. All the other cases of syntactical construction can be proven following the same proof pattern.

$$
X \to X' = \begin{cases}
M \to M' & , M \notin C, X = +1(M), X' = +1(M') & (i) \\
1 & , X = +1(n_i), X' = n_{(i+1)} & (ii) \\
M \to M' & , M \notin C, X = -1(M), X' = -1(M') & (iii) \\
1 & , X = -1(n_i), X' = n_{(i-1)}, i \geqslant 1 & (iv) \\
1 & , X = -1(n_0), X' = n_0 & \\
M \to M' & , M \notin C, X = 0?(M), X' = 0?(M') & (vi) \\
1 & , X = 0?(n_0), X' = \dot{t} & (vii) \\
1 & , X = 0?(n_i), X' = \dot{f}, i \geqslant 1 & (viii) \\
M \to M' & , M \notin C, X = \mathit{if}(M, N', N''), X' = \mathit{if}(M', N', N'') & (ix) \\
1 & , X = \mathit{if}(\dot{t}, N', N''), X' = N' & (x) \\
1 & , X = \mathit{if}(\dot{f}, N', N''), X' = N'' & (xi) \\
M \to M' & , M \neq (\lambda v.B), X = MN, X = M'N & (xii) \\
1 & , X = (\lambda v.B)N, X' = B[v := N] & (xiii) \\
1 & , X = \mu M, X' = M(\mu M) & (xiv) \\
0.5 & , M \neq N, X = M|N, X' = M & (xv) \\
0.5 & , M \neq N, X = M|N, X' = N & (xvi) \\
1 & , X = M|M, X' = M & (xvii) \\
1 & , X \in C, X = X' & (xviii) \\
1 & , X = \lambda v.B, X = X' & (xix) \\
0 & , \mathit{else} & (xx)
\end{cases}
\tag{3.65}
$$

In Case of $X \in C$: In case X is a constant we can distinguish two cases, i.e., the case that $X = X'$ and the case that $X \neq X'$. Let us consider the case that $X = X'$. We start with the observation that the rule set Eqn. (3.62-i) contains the rule $\emptyset \vdash \langle X, X, 1\rangle$. Now, due to the definition of $\to$ we know that $\to$ is closed under the rule $\emptyset \vdash \langle X, X, 1\rangle$ and therefore $\langle X, X, 1\rangle \in \to$. Now, we need to prove that $\langle X, X, 0\rangle \notin \to$ and $\langle X, X, 0.5\rangle \notin \to$. We prove that $\langle X, X, 0\rangle \notin\to$ only. Let us consider the set $\to\backslash\langle X, X, 0\rangle$. We observe that $\to\backslash\langle X, X, 0\rangle$ is closed under the complete collection of rule sets Eqns. (3.54) through (3.64). This can be understood by walking through all of these rule sets and checking that none of the contained rules requires $\langle X, X, 0\rangle \in \to$. Therefore, actually we know that $\langle X, X, 0\rangle \notin\to$, because $\langle X, X, 0\rangle \in\to$ would contradict that fact that $\rightarrow$ is the least relation that is closed under these rule sets. □

Note that in Lemma 3.3 we have used and, henceforth, we will use $M \rightarrow N$ to denote the result of applying the one-step semantics as a function to a pair of given closed terms M and N, i.e., we define

$$(M \rightarrow N) =_{DEF} \rightarrow(M, N) \tag{3.66}$$

This means that we use $M \rightarrow N$ to denote the uniquely given probability of the immediate transition from M to N, i.e., we have that

$$(M \rightarrow N) = i \quad \textbf{iff} \quad M \xrightarrow{i} N \tag{3.67}$$

Similarly, we use $M \rightarrow N$ to denote the matrix element in the M-th row and N-th column instead of the usual notation $\rightarrow_{MN}$. As expressed by Corollary 3.4 it turns out that the one-step semantics forms a probabilistic matrix.

Corollary 3.4 (Probabilistic Matrix) *The one-step semantics forms a probabilistic matrix, i.e., for each closed lambda term M we have*

$$\Big(\sum_{N \in \Lambda_\emptyset} M \rightarrow N\Big) = 1 \tag{3.68}$$

Proof. The Lemma follows immediately from Lemma 3.3 by a structural induction over term tuples M and N. In particular, each row contains either exactly one element having value 1 or two elements with both having value 0.5, whereas all other elements are zero. □

3.3.2 Markov Chain Semantics

Now, based on Corollary 2.20 and Corollary 3.4 we can assume the existence of a Markov chain $S = (S_n)_{n \geqslant 0}$ with the defined one-step semantics as transition matrix, i.e.,

$$S = \mathfrak{M}(\rightarrow) \tag{3.69}$$

Each random variable S_n models the probabilities that a reduction reaches given terms after n steps based on the initial distribution ι. Note that the concrete initial distribution ι is not relevant for our semantics. It is determined by the arbitrarily choosen, but fixed Markov chain $\mathfrak{M}(\rightarrow)$ and will factor out in the definition of reduction probabilities based on hitting probabilities.

Henceforth, we use S to denote the Markov chain $\mathfrak{M}(\rightarrow)$. Nevertheless, we feel free to switch between the usage of S and $\mathfrak{M}(\rightarrow)$ in the sequel. Typically, will use $\mathfrak{M}(\rightarrow)$ whenever we want to bring to mind the construction aspect of the Markov chain semantics. We say that $S = \mathfrak{M}(\rightarrow)$ is the reduction chain of the probabilistic lambda calculus. We also say that $S = \mathfrak{M}(\rightarrow)$ is the Markov chain, or even that it is the Markov chain semantics of the probabilistic lambda calculus.

Now, we will exploit the notion of hitting probability given in Def. 2.23 to define the reduction probability $M \Rightarrow N$. The reduction probability is the central notion of the operational semantics. The reduction probability $M \Rightarrow N$ is the probability that the reduction of a term M ever hits the term N, i.e., that the Markov chain S ever hits N conditional on $S_0 = M$.

Definition 3.5 (Reduction Probability) *Given closed terms M and N we define the reduction probability $M \Rightarrow N$ as follows:*

$$M \Rightarrow N \quad =_{DEF} \quad \eta_{\mathfrak{M}(\rightarrow)}\langle M, N\rangle \tag{3.70}$$

Given terms M and N, we also say that M reduces to N with probability $M \Rightarrow N$. As usual, we are particularly interested in programs, i.e., closed ground terms. Consequently, we are particularly interested in reduction probabilities $M \Rightarrow c$ from programs M to constants $c \in C$. We say M evaluates to c with probability $M \Rightarrow c$ in these cases. Also, we call the reduction probability $M \Rightarrow c$ an evaluation probability in such a case. Furthermore, we speak of the evaluation semantics of a program. Given a program M, its evaluation semantics is the function that assigns the respective evaluation probability to each constant.

Definition 3.6 (Evaluation Semantics) *Given a program M, its evaluation semantics $M \Rightarrow \_ : C \rightarrow [0,1]$ is naturally given by $c \in C \mapsto (M \Rightarrow c)$.*

There might be several different possibilities to construct an operational semantics for the probabilistic lambda calculus and the Markov chain semantics is one possibility, however, in this book we prefer to use operational semantics and Markov chain semantics as synonyms. The transition matrix $M \stackrel{i}{\rightarrow} N$, the Markov chain $S = \mathfrak{M}(\rightarrow)$ and the reduction probability $M \Rightarrow N$ are the essential ingredients of the Markov chain semantics.

Although it is fair to say that the Markov chain semantics is a typical operational semantics, operational semantics is a more general notion. It stands for a certain style of semantics that is distinguished from other styles of semantics such as axiomatic semantics or denotational semantics. It is typical that an operational semantics is close to operational concepts such as program execution, program reduction or program transformation. However, we are not restricted in the range of operational phenomena that we want to study. Particularly interesting operational aspects are found in the fields of algorithmic complexity and termination behavior. For example, based on the Markov chain semantics, it is possible to define the operational concept of mean reduction length. The mean reduction length yields the average number of steps $\varnothing(M \Rightarrow N)$ needed to reduce a program M to another program N.

Definition 3.7 (Mean Reduction Length) *Given closed terms M and N we define the mean reduction length $\varnothing(M \Rightarrow N)$ of the reduction from M to N as follows:*

$$\varnothing(M \Rightarrow N) = |\eta_{\mathfrak{M}(\rightarrow)}\langle M, N\rangle| \tag{3.71}$$

See how $\varnothing(M \Rightarrow N)$ is defined as the expected value of steps needed to reach N from M by Eqn. (3.71); see also the definition of mean hitting time in Def. 2.32. Programs that are equal with respect to their evaluation semantics might nevertheless be different with respect to other operational aspects. We will discuss this further in the next chapter on termination behavior.

Given the Markov chain semantics, we see that the choice operator is non-associative, i.e., in general we have that $(L|M)|N \Rightarrow c \neq L|(M|N) \Rightarrow c$.

Corollary 3.8 (Non-Associativity of the Choice Operator) *The evaluation semantics of the choice operator is non-associative.*

Proof. Take the programs $M_l = (1|2)|3$ and $M_r = 1|(2|3)$ as a counter-example. On the one hand we have that $(M_l \Rightarrow 1) = 0.25$, $(M_l \Rightarrow 2) = 0.25$ and $(M_l \Rightarrow 3) = 0.5$. On the other hand we have that $(M_r \Rightarrow 1) = 0.5$, $(M_r \Rightarrow 2) = 0.25$ and $(M_r \Rightarrow 3) = 0.25$. □

3.3.3 An Example Probabilistic Program

Let us see how the operational semantics works for an example term. Let us consider the following term m:

$$m =_{DEF} \mu\lambda f.\lambda x.(\ x \mid f(+1(x))\ ) \tag{3.72}$$

The term m has the functional type $num \rightarrow num$. Let us see what happens it m is applied to one of the number constants n_i. You can easily check the following. For each $j \geqslant i$, the term $m\, n_i$ evaluates to the constant n_j with probability 0.5^{1+j-i}, whereas for each $j < i$, it evaluates to the constant n_j with probability zero, i.e.,

$$m\, n_i \Rightarrow n_j = \begin{cases} 0 & ,j < i \\ 0.5^{1+j-i} & ,j \geqslant i \end{cases} \tag{3.73}$$

For example, consider the application of m to the first three numbers n_0, n_1 and n_2, which we denote by their usual number symbols in the example:

$(m\,0 \Rightarrow 0) = 0.5$	$(m\,1 \Rightarrow 0) = 0$	$(m\,2 \Rightarrow 0) = 0$
$(m\,0 \Rightarrow 1) = 0.25$	$(m\,1 \Rightarrow 1) = 0.5$	$(m\,2 \Rightarrow 1) = 0$
$(m\,0 \Rightarrow 2) = 0.125$	$(m\,1 \Rightarrow 2) = 0.25$	$(m\,2 \Rightarrow 2) = 0.5$
$(m\,0 \Rightarrow 3) = 0.0625$	$(m\,1 \Rightarrow 3) = 0.125$	$(m\,2 \Rightarrow 3) = 0.25$
$(m\,0 \Rightarrow 4) = 0.03125$	$(m\,1 \Rightarrow 4) = 0.0625$	$(m\,2 \Rightarrow 4) = 0.125$
$\cdots$	$\cdots$	$\cdots$

It is interesting to have a look at the termination behavior of programs of the form $m\,n_i$. The probability that a program $m\,n_i$ hits any of the number constants sums up to one as you can see in the following equation for any arbitrary but fixed n_i:

$$\eta_S\langle m\,n_i, C\rangle = \sum_{j=i}^{\infty} m\,n_i \Rightarrow n_j = \sum_{j=i}^{\infty} 0.5^{1+j-i} = \sum_{j=1}^{\infty} 0.5^j = 1 \tag{3.74}$$

Later, in Chap. 4, we will call the probability that a program ever reaches a constant the termination degree of the program; compare with Def. 4.1. We will see how to exploit linear algebra to argue about the termination degree of programs.

Of course, we can apply m not only to constants but to arbitrary terms. Let us consider what happens if we apply it to the term $n_0|n_1$. Here, the probability that $m(n_0|n_1)$ reduces to any of the number constants n_j equals the sum of $0.5 \cdot (m\,n_0 \Rightarrow n_j)$ and $0.5 \cdot (m\,n_1 \Rightarrow n_j)$, i.e., we have:

$$\begin{array}{l} (m\,(0|1) \Rightarrow 0) = 0.25 \\ (m\,(0|1) \Rightarrow 1) = 0.375 \\ (m\,(0|1) \Rightarrow 2) = 0.1875 \\ (m\,(0|1) \Rightarrow 3) = 0.09375 \\ \quad \ldots \end{array}$$

Now, let us analyze the probability $m\,M \Rightarrow n_j$ in its most general form, i.e., for arbitrary terms M and number constants n_j. In general, we have the following:

$$(m\,M \Rightarrow n_j) \;=\; \sum_{i=0}^{j} 0.5^{1+i} \cdot (M \Rightarrow n_{j-i}) \tag{3.75}$$

Actually, via a couple of steps, it can be seen that the probabilities for the special case $n_0|n_1$ above can be derived as instances of Eqn. (3.75). Now, let us explain how to come up with such a result as Eqn. (3.75). Note that the following is an informal argumentation, i.e., not a proof. Argumentations like the one conducted here can be turned into formal proofs by establishing structural inductions or exploiting linear algebra. First, we can conduct a one-step decomposition of hitting probabilities as described by Eqn. (2.35). Now, due to the definition of m in Eqn. (3.72) we can see that $(m\,M \Rightarrow n_j)$ equals the following:

$$0.5 \cdot (M \Rightarrow n_j) \;+\; 0.5 \cdot\; (m(+1(M)) \Rightarrow n_j) \tag{3.76}$$

Now, we can unfold Eqn. (3.76) further, *ad infinitum*, so that we can see that $(m\,M \Rightarrow n_j)$ equals the following for each arbitrary number k:

$$\Big(\sum_{i=0}^{k} 0.5^{1+i} \cdot +1^i(M) \Rightarrow n_j\Big) + \Big(0.5^{k+1} \cdot m(+1^{k+1}(M)) \Rightarrow n_j\Big) \tag{3.77}$$

Note, that terms of the form $+1^i(M)$ in Eqn. (3.77) are an ad hoc, informal notation. We use it to denote the i-times syntactical application of $+1$ to the term $(m\,M)$, i.e., $+1(+1(+1(...mM)))$ i times. Next, we can exploit the rule sets of (3.54) for addition terms $+1(N)$ in our argumentation. First, it can be understood by (3.54-ii), i times, that a term of the form $+1^i(n_{j-i})$ leads to the following one-step transitions:

$$+1^i(n_{j-i}) \xrightarrow{1} +1^{i-1}(n_{j-i+1}) \xrightarrow{1} +1^{i-2}(n_{j-i+2}) \xrightarrow{1} \cdots \xrightarrow{1} +1(n_{j-1}) \xrightarrow{1} n_j \tag{3.78}$$

Note that the probability of each of the transitions in Eqn. (3.78) is 100 per cent. Similarly, it can be understood that for all other $n_k \neq n_{j-i}$ the term $+1^i(n_k)$ must lead to transitions of the following form:

$$+1^i(n_k) \xrightarrow{1} +1^{i-1}(n_{k+1}) \xrightarrow{1} \cdots \xrightarrow{1} n_{k'} \xrightarrow{0} n_j \tag{3.79}$$

Have a look at the last step in Eqn. (3.79). Due to $k \neq j - i$ we know that $k' \neq j - 1$. Therefore, due to the rule scheme (3.63) we know that $n_{k'} \xrightarrow{0} n_j$, i.e., there is a zero per cent probability to step from $n_{k'}$ to n_j. Altogether, based on Eqns. (3.78) and (3.79) we know that $(+1^i(n_{j-i}) \Rightarrow n_j) = 1$ and $(+1^i(n_k) \Rightarrow n_j) = 0$ for all $n_k \neq n_{j-i}$. Based on that and the rule set (3.54-i) it can be understood that for all terms $N \in \Lambda_P$ we have the following evaluation semantics of the n-fold addition operator:

$$+1^i(N) \Rightarrow n_j = \begin{cases} N \Rightarrow n_{j-i} & , i \leqslant j \\ 0 & , \mathit{else} \end{cases} \tag{3.80}$$

Once more note that all of the argumentation given here is informal. We do not use formal inductions. Also we do not aim to make explicit the complete case distinctions. This will change in formal proofs, e.g., when we investigate the semantic correspondence between operational semantics and denotational semantics in Sect. 5.3. Now, given Eqn. (3.80) we have that Eqn. (3.77) equals

$$\Big(\sum_{i=0}^{k} 0.5^{1+i} \cdot M \Rightarrow n_{j-i}\Big) + \Big(0.5^{k+1} \cdot +1^{k+1}(m\,M) \Rightarrow n_j\Big) \tag{3.81}$$

Now, given the case distinction in Eqn. (3.80), we can further narrow the range of the stepping index i in Eqn. (3.81) to $0 \leqslant i \leqslant j$, which yields Eqn. (3.75).

3.3.4 Remarks On Multiset Semantics

In the definition of the one-step pre-semantics for the nondeterministic choice, i.e., for terms of the form $M|N$, we have explicitly distinguished between the case $M \neq N$ in rules (3.61-i) and (3.61-ii) and the case $M = N$ in rule (3.61-iii). Intuitively, it is an option not to distinguish between these two cases in the definition of the one-step pre-semantics and to replace all of the rules in Eqn. (3.61) simply by

$$(i)\frac{}{M|N \xrightarrow{0.5} M} \quad (ii)\frac{}{M|N \xrightarrow{0.5} N} \tag{3.82}$$

However, the formalization given with Eqn. (3.82) would only work with a non-standard understanding of inference rules because in case $M = N$ not one but two labeled transitions of the form $\xrightarrow{0.5}$ would be introduced between $M|M$ and M. For this purpose we would have to use a non-standard definition of labeled transition systems that allows for multiple transitions between nodes, e.g., by exploiting the notion of multisets [29]. However, ultimately we are interested in exploiting the one-step semantics as a probability matrix. Therefore, we would have to flatten multiple transitions of the form $\xrightarrow{0.5}$ to a single matix entry of value 1 between terms of the form $M|M$ and M. For these reasons, we have chosen the formalization given by Eqn. (3.61). The given formalization is a completely adequate description of the intended operational semantics, because if a reduction has reached a term of the form $M|M$ there is one possible and at the same time necessary next reduction to M.

3.3.5 Remarks On Choices with Arbitrary Probabilities

Note that the probabilistic choice $M|N$ as defined by Eqn. (3.61) has a fixed, somehow arbitrary probability of 0.5 for each possible choice. It would be an alternative to introduce a general choice construct of the form $M_i|N$ for $0 \leqslant i \leqslant 1$ that allows for the specification of an arbitrary probability for concrete choices by the modeler with appropriate rules such as

$$(i)\frac{M \neq N}{M_i|N \xrightarrow{i} M} \quad (ii)\frac{M \neq N}{M_i|N \xrightarrow{1-i} N} \quad (iii)\frac{}{M_i|M \xrightarrow{1} M} \tag{3.83}$$

Yet another possibility would be to work with an even more general choice construct $M_i|_jN$ with $0 \leqslant i + j \leqslant 1$ that allows for the specification of a probability for each direction of the choice with appropriate rules such as

$$(i)\frac{M \neq N}{M_i|_jN \xrightarrow{i} M} \quad (ii)\frac{M \neq N}{M_i|_jN \xrightarrow{j} N} \quad (iii)\frac{}{M_i|_jM \xrightarrow{i+j} M} \tag{3.84}$$

The point is that more general choice constructs such as Eqns. (3.83) and (3.84) do not add much to the expressive power. This means that the choice construct $M|N$ with fixed probability is able to simulate such more general choice constructs. In [139] it has been shown that a probabilistic choice with fixed probability is sufficient to implement a generalized probabilistic choice.

The algorithm in [139] is also used in [185] where it is turned into an imperative version and proven correct with the axiomatic reasoning framework of [185]. Let us sketch a program $M\,{}_i|^{\Lambda}\,N$ that is intended to simulate the program $M_i|N$ from Eqn. (3.83) as follows:

$$
\begin{aligned}
M\,{}_i|^{\Lambda}\,N \equiv (\; & \mu\lambda C.\lambda p.\lambda d. \\
& \quad \mathit{if}\,(p = i)\ \mathit{then}\ N\ \mathit{else} \\
& \quad \mathit{if}\,(p + 0.5^d \leqslant i)\ \mathit{then} \\
& \quad\quad M \mid (C\ (p + 0.5^d)\ (d+1)) \\
& \quad \mathit{else} \\
& \quad\quad N \mid (C\ p\ (d+1)) \\
)\; & 0\ 1
\end{aligned}
\tag{3.85}
$$

First, it has to be noted that $M\,{}_i|^{\Lambda}\,N$ is not yet a term of Λ, it is rather a sketch. This is so, because we make use of rational number arithmetic in $M\,{}_i|^{\Lambda}\,N$ as if it were built into Λ. For example, we have a constant 0.5 and the operation 0.5^d, i.e., a power function. But this should not concern us too much here. Obviously, it would be possible to extend the probabilistic lambda calculus with appropriate types, variables, constants and operators. However, it is also possible to represent rational arithmetic by appropriate terms without extending Λ. For example, we can represent each rational number $q = n/d$ as a term $\mathtt{q} =_{DEF} \lambda s.\mathit{if}(s, n, d)$. See how $\mathtt{q}$ packs the numerator n and the denominator d together and allows access to them via the selector s. Now, appropriate operations can be programmed. For example, addition of rational numbers can be programmed as follows:

$$
\begin{aligned}
\mathtt{plus_Q} \equiv \lambda x.\lambda y.\lambda s.\mathit{if}\,(s,\; & \\
& (\mathtt{plus_N}(\mathtt{mult_N}\,(x\,\dot{t})(y\,\dot{f}))(\mathtt{mult_N}\,(x\,\dot{f})(y\,\dot{t}))), \\
& (\mathtt{mult_N}\,(x\,\dot{f})(y\,\dot{f})) \\
)\; &
\end{aligned}
\tag{3.86}
$$

The operation $\mathtt{plus_Q}$ in Eqn. (3.86) is again only a sketch, because it relies on the existence of properly programmed operators for multiplication and addition of natural numbers. Fortunately, natural numbers are already represented in Λ by the type C_{num}. We do not want to delve into this further here, we just wanted to show that it is, in principle, possible to represent rational number arithmetic in Λ as is. Therefore, we can proceed with our discussion of $M\,{}_i|^{\Lambda}\,N$ in Eqn. (3.85).

See, how $M\,{}_i|^{\Lambda}\,N$ works. It realizes an interval nesting around the targeted probability i. The program $M\,{}_i|^{\Lambda}\,N$ recursively calls itself and upon each recursive call it might give M or N the chance to execute. In d the program keeps track of the number of recursive calls already executed, i.e., the recursion depth. In p the program keeps track of the chance already granted to term M during the course of the execution. Think of p as an accumulated probability. Now, it is p that steers which term, i.e., either M or N, gains the chance to execute. It can be shown that the probability of $M\,{}_i|^{\Lambda}\,N$ ending up in M is

i, whereas the probability of $M\ {}_i|^{\Lambda}\ N$ ending up in N is $1-i$. This is exactly what we wanted to achieve with $M\ {}_i|^{\Lambda}\ N$. In order to prove a program like $M\ {}_i|^{\Lambda}\ N$ correct, the apparatus of the developed Markov chain semantics can be exploited.

Albeit it is possible to simulate $M_i|N$ by $M\ {}_i|^{\Lambda}\ N$, the two programs are still different with respect to other operational aspects such as run-time behavior and termination behavior. We have not yet developed the appropriate terminology and apparatus for the systematic investigation of such aspects and refer to Chap. 4 for this purpose.

In Eqn. (3.83) we have assumed that the probability specification is an arbitrary but fixed number $0 \leqslant i \leqslant 1$. A further interesting idea would be to allow terms as probability specifications i. This way, it would be possible to let probabilities of choice evolve dynamically. Again the operational semantics for such a programming language construct can be given in a straightforward manner. We just need some further reduction rules like these:

$$(i)\frac{c_i \in C \quad M \neq N}{M_{c_i}|N \xrightarrow{i} M} \qquad (ii)\frac{c_i \in C \quad M \neq N}{M_{c_i}|N \xrightarrow{1-i} N} \qquad (iii)\frac{}{M_Q|M \xrightarrow{1} M} \tag{3.87}$$

$$(iv)\frac{P \notin C \quad P \xrightarrow{j} Q \quad M \neq N}{M_P|N \xrightarrow{j} M_Q|N} \tag{3.88}$$

If the probabilistic choice allows us to use real numbers from $[0,1]$ in the specification of arbitrary probabilities, the whole programming language consists of an uncountable number of terms. Actually, this is a minor issue, because the set of terms reachable from a given starting program is, of course, always countable and for the purpose of semantics it is therefore always possible to consider a countable subset of the whole language.

3.3.6 Remarks On Probabilistic Choice at Higher Types

We introduced probabilistic choice at higher types; compare with rule (3.13). For example, we allow for a term such as $(\lambda x_{num}.x)\,|\,(\lambda y_{num}.{+1}(y))$. Actually, we allow for all terms $M|N$ of some higher type t. We could do this differently and introduce probabilistic choice at the ground type only. Actually, in a sense this would not change the expressive power of our language.

First, we still have the conditional construct *if-then-else* that allows for branching to terms of arbitrary type, i.e., also of some higher type. Given two terms M and N of some higher type t, let us assume that we are not allowed to express the probabilistic choice directly as $T \equiv M|N$. Still, we can realize the probabilistic choice with the term $T' \equiv \mathit{if}(0|1)\,\mathit{then}\,M\,\mathit{else}\,N$ in which the term $(0|1)$ plays the role of a coin flip. Actually, the terms T and T' behave the same, i.e., they are operationally equivalent. Two terms M and N are said to be operationally equivalent if substituting them into an arbitrary context

$K[\_]$ of appropriate type always leads to terms $K[M]$ and $K[N]$ that have the same evaluation semantics, i.e., $(K[M] \Rightarrow C) = (K[N] \Rightarrow C)$; compare with Def. 3.6. We do not provide a proof for this here.

Second, probabilistic choice for higher-typed terms M and N can be simulated anyway, i.e., without the need to exploit the conditional construct. Given again that the terms M and N have type t, we know by Lemma 3.1 that t has a top-level functional structure $t = (t_n \to (\ldots(t_2 \to (t_1 \to g))...))$ for n types $t_i \in T$ and a ground type $g \in T_g$. Now, we can write the following term for n fresh variables x_{t_i}, i.e., variables that are not free in either M or N:

$$T'' \equiv \lambda x_{t_1}.\lambda x_{t_2}.\ldots.\lambda x_{t_n}.((M x_{t_1} \cdots x_{t_n}) \mid (N x_{t_1} \cdots x_{t_n})) \tag{3.89}$$

Now, we have again that T'' from Eqn. (3.89) is operationally equivalent to $T = (M|N)$. Again, we do not provide a proof for this, but it should be easy to see that it is the case, because $T = T''$ is a kind of η-equivalence. Remember that η-equivalence [18] expresses extensionality, i.e., the fact that each term M converts to $\lambda x.(Mx)$ as long as x is not free in M. Therefore, we can see that T and the term $T''' \equiv \lambda x_{t_1}.\lambda x_{t_2}.\ldots.\lambda x_{t_n}((M|N)x_{t_1} \cdots x_{t_n})$ are actually operationally equivalent. Now, it should be easy to see that also T'' and T are operationally equivalent.

This means we could restrict the probabilistic choice to ground types without loss of operational expressiveness, in any case, i.e., even if our conditional construct did not allow for higher-typed branches. However, having the probabilistic choice at higher type fits together with the crucial notion of functional programming languages, i.e., that functions are first-class citizens. In functional programming languages functions are fundamental building blocks that can be passed around as parameters and used as data in data structures. The design of functional algorithms, functional data structures [213], programs and programming systems heavily relies on that notion of functions as first-class citizens. For example, the widely known functional *map-reduce* programming pattern gathers functions as objects in a data structure.

The typed lambda calculus is a maximally reductionist functional programming language. As such, it is interesting to investigate the semantics of the typed lambda calculus. The typed lambda calculus is, of course, also a model of computation. As such it is only one option among others for studying probabilistic computation. For example, Dana Scott investigates probabilistic choice for the untyped lambda calculus in [242] and Dal Lago and Zuppiroli characterize probabilistic computation in the style of Kleene's partial recursive functions in [59, 60]. A similar question is whether treating probabilistic choice at higher types or restricting it to ground types can be asked with respect to any programming primitive or building block, e.g., for the conditional *if-then-else*. In Plotkin's original version of PCF [223], the *if-then-else* construct is restricted to ground types, i.e., terms $\mathit{if}(C, M, N)$ can only be formed for terms M and N of ground type. In the book by Gunther [118], a standard textbook on semantics of programming languages, PCF comes in a version that allows for higher-typed conditionals, i.e., so that the branches M and N might have higher type. Now, the same discussion as conducted for the probabilistic choice applies. Again, we can rewrite a higher-typed conditional term $R \equiv \mathit{if}(C, M, N)$ by an operationally equivalent term

$$R' \equiv \lambda x_{t_1} \ldots . \lambda x_{t_n} . \mathit{if}(C, (M x_{t_1} \cdots x_{t_n}), (N x_{t_1} \cdots x_{t_n}))$$

for an approriate vector of variables such that $M x_{t_1} \cdots x_{t_n}$ and $N x_{t_1} \cdots x_{t_n}$ have ground type. We can even go further. Yet, the conditional construct is not in its most reductionist form. We can further restrict the branches of the conditional construct from arbitrarily terms of ground type to variables of ground type. Still, we can write a term

$$R'' \equiv \lambda x_{t_1} \ldots . \lambda x_{t_n} . (\lambda\, l.\, \lambda\, r. \mathit{if}(C, l, r))(M x_{t_1} \cdots x_{t_n})(N x_{t_1} \cdots x_{t_n})$$

for ground-type variables l, r and further variables $x_{t_1} \cdots x_{t_n}$ that are all not free in M and N so that R'' is operationally equivalent to R. Now, with $\lambda\, l.\, \lambda\, r. \mathit{if}(C, l, r)$ in R'' we have arrived at conditional terms that are purely combinatorical, i.e., just select between the arguments that they are given.

With respect to the elaboration of the operational semantics, the introduction of the probabilistic choice at higher types comes at no price. It is not relevant for the structures and techniques that we establish in the operational semantics, whether we introduce the probabilistic choice generally for all types or restrict it to ground types. At the same time the general introduction for all types is more straightforward and somehow more natural, i.e., in better accordance with the notion of functions as first-class citizens. Actually, with respect to the elaboration of the denotational semantics the generally typed probabilistic choice comes at a price, albeit not a very high price. We need to define functional versions of the vector space operations, see Sect. 5.1.4, to be used in the semantic equations. On the other hand, fortunately, that is all. Even with a generally typed probabilistic choice we need to introduce distributions only at the level of flat domains in our semantics. With the functional vector space operations from Sect. 5.1.4 we can consume or flatten probabilistic choice at higher types immediately whenever it occurs in semantic equations. The functional vector space operations semantically replay the pattern that is shown in Eqn. (3.89).

3.3.7 Remarks On Call-by-Value

Reduction strategies constitute a wide field. Not only do they affect the semantics of a programming language and, therefore, program and program system design, but they are crucial in programming language implementation, i.e., when it comes to non-functional aspects such as performance; see, e.g., the book of Simon Peyton Jones [217]. We cannot delve into these issues here and only glance at the difference between call-by-value and call-by-name [220] where we are interested only in direct semantical differences. The operational semantics that we have introduced is a call-by-name semantics. In deterministic lambda calculi call-by-value and call-by-name differ only with respect to termination behavior. Programs that terminate under both of the reduction strategies evaluate to the same value. Some programs that terminate under call-by-name might not terminate under call-by-value. In a probabilistic calculus, the difference between the two reduction strategies is more fine grained. Let us have a look at the following program:

$$T \equiv \underbrace{(\lambda f.f(f(0)))}_{M} \underbrace{(\, \underbrace{(\lambda x. +1(x))}_{N_1} \mid \underbrace{(\lambda x. +1(+1(x)))}_{N_2} \,)}_{N} \tag{3.90}$$

Under the call-by-name semantics that we have introduced, the program T reduces to value 2 with a 25% probability, to value 3 with a 50% probability and to value 4 again with a 25% probability. Under call-by-name the argument term N is substituted into the program body of M as a whole. Under call-by-value it is decided earlier, before function application, whether N_1 or N_2 eventually can affect the reduction and the occurence of the term $+(+(+(0)))$ is pre-empted. Therefore under call-by-value the program T reduces to value 2 with a 50% probability and to value 4 also with a 50% probability.

With respect to the elaboration of the denotational semantics the difference between call-by-name and call-by-value matters. As we have already described in Sect. 3.3.6, in the case of call-by-name it suffices to introduce probabilistic distributions over flat domains. The semantics of the term N_1 is just a function that takes a natural number and yields a distribution over the natural numbers. It is not a distribution over this or any other kind of functions. The same goes for the term N_2. Then, the semantic version of the probabilistic choice merges the semantics of N_1 and N_2 with appropriate functional vector space operations. In such merging the information from where single probabilities stem vanishes, which does no harm. In the case with call-by-value this cannot work. In a denotational semantics with call-by-value, we must keep track of such information, so that it can be exploited in function application.

With respect to the elaboration of the operational semantics there is no essential difference between call-by-name and call-by-value. Actually, the investigation of termination behavior and its results in Chap. 4 hold independently of the chosen reduction strategy.

We complete these remarks by giving the one-step-semantics rules for the case of call-by-value. All that is needed is to replace (3.59-ii) by the following two rules:

$$i)\frac{N \in \Lambda_V}{(\lambda v.B)N \xrightarrow{1} B[v := N]} \qquad ii)\frac{N \notin \Lambda_V \quad N \xrightarrow{i} N'}{(\lambda v.B)N \xrightarrow{i} (\lambda v.B)N'} \tag{3.91}$$

The set Λ_V in Eqn. (3.91) is the set of values that consists, as usual, of constants and lambda abstractions; compare with Def. 3.41.

3.4 Important Readings

The work [232] of Saheb-Djahromi gives an operational and a denotational semantics to a version of the probabilistic lambda calculus. The host language is the higher-typed lambda calculus. Probabilistic choice is introduced at the level of ground types. The language allows for arbitrary real number probabilities as annotations to the probabilistic choice. The language allows for both call-by-name and a limited form of call-by-value. Two different kinds of lambda abstractions are introduced for the purpose of steering the evaluation. Call-by-value lambda abstractions are limited to variables of ground types. The operational semantics starts with a one-step semantics, which forms a probabilistic matrix. Reduction is not explained as the hitting probability of a Markov chain. Instead, on the basis of the probabilistic matrix, all finite evaluation sequences are defined inductively. Then, an evaluation semantics is defined as the limit of finite evaluation sequences. In this approach to operational probabilistic semantics the decomposition of reduction probabilities is the defining concept, i.e., it becomes part of an explicit construction and does not follow as a property of a Markov chain semantics; compare with Lemma 2.31 and Eqn. (2.35).

For predicate transformer semantics of probabilistic programming languages, see the work of Keimel et al. in [150]. The work describes predicate transformers for partial correctness semantics for a reductionist imperative programming language that contains primitives for both probabilistic and nondeterministic choice. Compare also the work of Tix, Keimel and Plotkin in [258]; see Sect. 5.4.

In [58] Dal Lago and Zorzi define operational semantics for probabilistic versions of the untyped lambda calculus, i.e., for both the call-by-name and the call-by-value case. For both cases, small-step and big-step evaluation semantics are provided. Here, both inductive and co-inductive [170] definitions are provided. The mutual relationships between the several operational semantics are investigated. The small-step semantics are shown to be equivalent to the respective big-step semantics. Also, standard continuation-passing-style transformations [220] between call-by-value and call-by-name semantics are generalized to the probabilistic case.

The seminal work on probabilistic bisimulation is the paper [174] by Larsen and Skou. In [62, 61] Dal Lago, Sangiorgi and Alberti define a notion of applicative bisimulation [2] for the probabilistic untyped lambda calculus. They investigate congruence and completeness of the resulting probabilistic applicative bisimulation. Congruence is shown and a counterexample to completeness is provided for the deterministic fragment as well as a further specifically refined notion of bisimulation. In [54, 55] Crubillé and Dal Lago proceed with the study of probabilistic applicative bisimulation for the probabilistic typed, call-by-value lambda calculus. It turns out that in this case, probabilistic applicative bisimulation exactly meets context equivalence.

In [59, 60] Dal Lago and Zuppiroli take a genuinely different approach to probabilistic computation. They characterize it from scratch in the style of Kleene recursive function definitions [151], without detour via deterministic computation. This way they can bridge more directly to recursion-theoretical and complexity-theoretical considerations; compare also with Dal Lago and Toldin [56] for an early reference and again later to [57].

Languages such as IBAL [218], Church [110] and Venture [180] are probabilistic programming languages, called rational programming languages in the case of IBAL [218]. Their genuine purpose is not to implement randomized algorithms but to model and generate stochastic distributions; compare also with Sect. 1.5. Appropriate formal semantics of such languages must model the evolution of stochastic distributions along the traces of probabilistic program executions. In [228], Ramsey and Pfeffer define a version of the untyped probabilistic lambda calculus that incorporates continuous distributions as computational atoms as well as a primitive to query these distributions. They give a monadic semantics to the resulting stochastic lambda calculus and provide a Haskell implementation for it. Compare also the work of Park in [214, 215]. In [35, 36] Borgström et al. give a big-step, trace-oriented semantics for a stochastic untyped lambda calculus, which serves as a reductionist model of rational programming languages. Similarly, in [250] Staton et al. provide both an operational and a monad-based denotational semantics for a simply typed probabilistic lambda calculus; compare also with [249].

4

Termination Behavior

In this chapter we study the termination behavior of the probabilistic lambda calculus. The degree of termination is the central concept in these investigations. Given a program M, its degree of termination is the probability that it ever hits a constant value. We also simply say termination degree for the degree of termination of a program. We also say that the termination degree of a program is its probability to terminate.

Definition 4.1 (Degree of Termination) *Given a program M, its degree of termination (termination degree) equals $\eta_S\langle M, C\rangle$, i.e., the probability that it ever hits a constant value.*

In the probabilistic lambda calculus, it is not sufficient any more to talk about the termination and non-termination of a program as exclusive opposites. Rather, we need to distinguish between a program and its program runs. A single program run either terminates or does not, whereas a program has a termination degree, which is the probability that the program executes a terminating program run. Furthermore, we might have programs with some non-terminating program runs that, however, terminate with a hundred percent probability. As with standard, non-probabilistic programs, termination is about the reachability of constants. A program run that eventually reaches a constant value is said to terminate, whereas a program run that never reaches a constant is said to be non-terminating. In the standard, non-probabilistic case there is a one-to-one correspondence between programs and program runs. Each program specifies exactly one program run. In the probabilistic case, there is a one-to-many correspondence between a program and its program runs. The standard execution of a probabilistic program results in one of its program runs.

In the next sections we will make precise notions such as program runs, terminating program runs, non-terminating program runs and further related notions. Based on that, we will investigate termination behavior. We will introduce the notions of bounded and unbounded termination and show how these are related to the existence of non-terminating program runs. We will identify a widened notion of termination, so-called path stoppability or p-stoppability, which allows us to determine the degree of termination of certain programs albeit they might have non-terminating program runs.

Please have a quick look at Tables 4.1 and 4.2. These tables give an overview of the most important terminology and results of the next sections. You might want to have a glance at the tables to gain a first impression about the upcoming material. In any case, please use Tables 4.1 and 4.2 as a reference for the following discussion and observations.

	$\eta_S\langle M, C\rangle = 1$	$\eta_S\langle M, C\rangle < 1$
$T(M) = P(M)$	*All program runs terminate. The program terminates with degree one.* The program terminates bounded.	–
$T(M) \subset P(M)$	*Not all of the program runs terminate. The program terminates with degree one.* The program terminates unbounded.	*Not all of the program runs terminate. The program terminates with degree less than one.* It is not determined whether the program terminates bounded or unbounded.

Table 4.1. Bounded and unbounded termination

<table>
<tr><th></th><th>$\eta_S\langle M, C\rangle = 1$</th><th>$\eta_S\langle M, C\rangle < 1$</th><th></th></tr>
<tr><td>$T(M) = P(M)$</td><td>All program runs terminate. The program terminates with degree one.</td><td>–</td><td rowspan="2">finite $Cover(M)$</td></tr>
<tr><td rowspan="2">$T(M) \subset P(M)$</td><td colspan="2">Not all of the program runs terminate. However, all program runs are p-stoppable, because the term cover of the program is finite. Program dovetailing is p-stoppable. The termination degree is computable, e.g., based on applications of Gauss-Jordan elimination.</td></tr>
<tr><td colspan="2">Not all of the program runs terminate. Not all of the program runs are p-stoppable. Program dovetailing is not p-stoppable. The termination degree is not computable.</td><td>infinite $Cover(M)$</td></tr>
</table>

Table 4.2. p-Stoppability and term covers

In Table 4.1 and Table 4.2 we use $P(M)$ to denote the set of program runs of M, whereas $T(M)$ stands for the set of terminating program runs of M. In principle, you can think of program runs and program executions as synonymous. It is only later, in Sect. 4.3, when we will distinguish between program runs and program execution for technical reasons. Obviously, with $T(M) = P(M)$ we indicate that all of the program runs of M terminate, whereas $T(M) \subset P(M)$ indicates that there exists a non-terminating program run. Table 4.1 summarizes how boundedness of program termination is related to the termination degree of a program. A program is said to be bounded if the execution lengths of all of its terminating program runs are capped by an upper bound. Otherwise, the program is said to be unbounded. We will define the notion of bounded and unbounded termination in Sect. 4.2. Next, Table 4.2 summarizes how finiteness of a term cover relates to the path stoppability of programs. With the term cover of a program M, or just cover of the program, we name the set of all terms that can be reached by any potential program execution starting from M. We also denote the cover of a program by $Cover(M)$. We investigate term covers and path stoppability Sect. 4.6.

In accordance with the degree of termination we say that $\eta_S\langle M, \overline{C}\rangle$ is the degree of non-termination of a program M, i.e., the probability that program M fails to terminate. Obviously, the degree of termination and the degree of non-termination of a program sum up to one, as expressed by Lemma 4.2.

Lemma 4.2 (Degrees of Termination and Non-Termination)
Given a program M we have that

$$\eta_S\langle M, \overline{C}\rangle = 1 - \eta_S\langle M, C\rangle \tag{4.1}$$

Proof. This follows from the fact that C and $\overline{C}$ are complements and the definitions of first hitting times and hitting probabilities in Defs. 2.22 and 2.23. First, we know that the sets $H_S(C) < \infty \wedge S_0 = M$ and $H_S(\overline{C}) < \infty \ \wedge \ S_0 = M$ are disjoint and sum up to $S_0 = M$. Next, we know that $\eta_S\langle M, C\rangle$ and $\eta_S\langle M, \overline{C}\rangle$ sum up to 1; also compare with Eqn. (2.24). □

The subject of investigation of this chapter is the termination behavior of the probabilistic typed lambda calculus. Here, we are not only interested in almost sure termination, i.e., termination degrees of one, but in the concepts of path stoppability and bounded termination, which are available for all termination degrees. For an investigation of termination behavior of imperative programming systems, see the work of Lehmann, Pnueli and Stavi in [176] and Hart, Sharir and Pnueli in [127, 128]. The subject of investigation in [176, 127, 128] is systems that consist of a finite number of concurrent processes, where each of the processes is a state-changing machine manipulating private and shared variables.

4.1 Introductory Examples of Termination Behavior

Before we start a rigorous investigation of termination behavior, we walk through a series of examples, and, for the time being, we use the concepts of program runs, terminating program runs, non-terminating program runs, termination, and degree of termination as informal, intuitive notions. It is the task of upcoming sections to make all of these notions precise and give a formal semantics to them. We begin with a very simple example, i.e., a program with termination degree 0.5. Given an arbitrary constant $n_i \in C_{num}$ we define the following program:

$$M_0 = n_i \mid \mu\lambda x.x \tag{4.2}$$

The program M_0 in Eqn. (4.2) allows for exactly two program runs, i.e., the one that terminates with n_i after one step and a second non-terminating program run:

$$n_i \mid \mu\lambda x.x \xrightarrow{0.5} n_i$$
$$n_i \mid \mu\lambda x.x \xrightarrow{0.5} \mu\lambda x.x \xrightarrow{1} (\lambda x.x)(\mu\lambda x.x) \xrightarrow{1} \mu\lambda x.x \xrightarrow{1} \cdots$$

Each of the program runs has probability 0.5 and therefore the program terminates with probability 0.5 and also diverges with probability 0.5. Note that the program runs that reaches the constant n_i does not stop after reaching n_i, i.e., it is an infinite program run that actually looks like this:

$$n_i \mid \mu\lambda x.x \xrightarrow{0.5} n_i \xrightarrow{1} n_i \xrightarrow{1} \cdots$$

When we say that a program reaches a constant, it reaches that constant necessarily after a finite number of steps. After it has reached the constant, it stabilizes. Therefore with respect to termination behavior, it makes sense to think of terminating program runs as finite executions, whereas it makes sense to think of non-terminating program runs as infinite executions. Henceforth, we will write down the infinite program runs that terminate, i.e., that reach a constant, as finite sequences of term reductions as we did in the above example. Either way, such notation in examples is merely informal notation. However, later, in Sect. 4.7 on program reduction trees we will turn the notion of finiteness vs. infiniteness of program runs into a concrete tree paths model.

Given an arbitrary but fixed constant $n_i \in C_{num}$, all of the following programs have the same evaluation semantics:

$$M_1 = -1(+1(n_i)) \mid n_i \tag{4.3}$$
$$M_2 = \mu\lambda x.(x \mid n_i) \tag{4.4}$$
$$M_3 = \mu\lambda x.(-1(+1(x)) \mid n_i) \tag{4.5}$$

The evaluation semantics of all the programs M_i in Eqns. (4.3) through (4.5) is the following:

$$M_i \Rightarrow c = \begin{cases} 1 & , c = n_i \\ 0 & , else \end{cases} \tag{4.6}$$

Consequently, the termination degree of all of the programs in Eqns. (4.3) through (4.5) is the same. Their termination degree equals one, i.e., there is a hundred percent probability that the program terminates. Although the evaluation semantics of all the programs is equal, they differ crucially in their operational behavior with respect to several aspects. These aspects are, basically, the average execution time, the question whether there exists a non-terminating program run or does not, the question whether the set of their reachable terms is finite and so forth. The first program M_1 reaches the result n_i always in a finite number of steps, i.e., either immediately after one step or otherwise after three steps:

$$\begin{aligned}
&-1(+1(n_i)) \mid n_i \xrightarrow{0.5} n_i \\
&-1(+1(n_i)) \mid n_i \xrightarrow{0.5} -1(+1(n_i)) \xrightarrow{1} -1(n_{i+1}) \xrightarrow{1} n_i
\end{aligned}$$

Program M_2 is different. It has a non-terminating program run ω_{M_2}, i.e., a program run that never reaches a constant value:

$$\omega_{M_2} = \mu\lambda x.(x|n_i) \xrightarrow{1} \cdots \xrightarrow{0.5} \mu\lambda x.(x|n_i) \xrightarrow{1} \cdots \xrightarrow{0.5} \mu\lambda x.(x|n_i) \xrightarrow{1} \cdots$$

The program run ω_{M_2}, taken as an event $\{\omega_{M_2}\}$, has a zero percent probability, i.e., $\mathsf{P}_{M_2}(\{\omega_{M_2}\}) = 0$. However, it is the only non-terminating program run of M_2 so we know that the termination degree of M_2 remains one. There is exactly one value, i.e., n_i, that can be reached by M_2. However, there are infinitely many program runs τ_i to reach the result that take the following form for each number i:

$$\tau_i = \underbrace{\mu\lambda x.(x|n_i) \xrightarrow{1} \cdots \xrightarrow{0.5} \mu\lambda x.(x|n_i)}_{i\ times} \xrightarrow{1} \cdots \xrightarrow{0.5} n_i \xrightarrow{1} n_i \cdots$$

Informally, a program run τ_i is a program run that loops i times before it takes its execution branch to the final result n_i. Altogether the program runs τ_i have a probability of one, because the probabilities of the individual program runs τ_i sum up to $0.5+0.25+0.125+\cdots$ which equals one, i.e., $\mathsf{P}_{M_2}(\{\tau_i \mid i \in \mathbb{N}_0\}) = 1$. Fortunately, we do not have to rely on such informal argumentation, because we have the Markov chain semantics to hand. Note that $\mathsf{P}_{M_2}(\{\tau_i \mid i \in \mathbb{N}_0\})$ is exactly the reduction probability $M_2 \Rightarrow n_i$. Now, we know that a finite number of terms are reachable by program executions of M_2, i.e., the term M_2 itself, $(\lambda x.(x|n_i))M_2$, $n_i|M_2$, and n_i. Therefore, and because of the Markov chain semantics and Theorem 2.29, we have that the corresponding vector of reduction probabilities is the solution of the following equation system:

$$\begin{aligned}
[M_2 \Rightarrow n_i] &= [(\lambda x.(x|n_i))M_2 \Rightarrow n_i] \\
[(\lambda x.(x|n_i))M_2 \Rightarrow n_i] &= [n_i|M_2 \Rightarrow n_i] \\
[n_i|M_2 \Rightarrow n_i] &= 0.5 \cdot [n_i \Rightarrow n_i] + 0.5 \cdot [M_2 \Rightarrow n_i] \\
[n_i \Rightarrow n_i] &= 1
\end{aligned} \tag{4.7}$$

Note that each term of the form $M \Rightarrow N$ in the above equation system Eqn. (4.7) stands for a single variable. We use brackets to denote these variables by $[M \Rightarrow N]$ in the equation system, however, we do this only in order to improve readability. Solving the equation system yields, among other things, that $M_2 \Rightarrow n_i$ equals one. To see this, just substitute variables from bottom to top in the equation system Eqn. (4.7); which simply yields

$$[M_2 \Rightarrow n_i] = 0.5 + 0.5 \cdot [M_2 \Rightarrow n_i] \tag{4.8}$$

$$[(\lambda x.n_i|x)M_2 \Rightarrow n_i] = 0.5 + 0.5 \cdot [M_2 \Rightarrow n_i] \tag{4.9}$$

$$[n_i|M_2 \Rightarrow n_i] = 0.5 + 0.5 \cdot [M_2 \Rightarrow n_i] \tag{4.10}$$

$$[n_i \Rightarrow n_i] = 1 \tag{4.11}$$

Now, solving Eqn. (4.8) yields a probability of one for $M_2 \Rightarrow n_i$.

Now, the difference between M_1 and M_2 is neither in the probability of reducing the program to the result, nor in the termination degree. The difference lies in the fact that M_2 has a non-terminating program run. Therefore, any number of steps may be needed to eventually arrive at the final result when M_2 is executed. This is different from the case of M_1. There, there exists a maximum number of steps needed to yield the result. Now, we can say that the termination of M_1 is bounded, whereas the termination of M_2 is unbounded.

We have said that program M_2 is unbounded, i.e., we cannot give a boundary for the number of steps it executes before it terminates. However, at least it is possible to determine the expected average number of steps needed by M_2 to yield the result value n_i. Again based on the Markov chain semantics, we know that the vector of average reduction lengths is the least solution to an equation system, which in case of M_2 yields the following:

$$\begin{aligned}
\varnothing(M_2 \Rightarrow n_i) &= 1 + \varnothing((\lambda x.(x|n_i))M_2 \Rightarrow n_i) \\
\varnothing((\lambda x.(x|n_i))M_2 \Rightarrow n_i) &= 1 + \varnothing(n_i|M_2 \Rightarrow n_i) \\
\varnothing(n_i|M_2 \Rightarrow n_i) &= 1 + 0.5 \cdot \varnothing(n_i \Rightarrow n_i) + 0.5 \cdot \varnothing(M_2 \Rightarrow n_i) \\
\varnothing(n_i \Rightarrow n_i) &= 0
\end{aligned} \tag{4.12}$$

Now, in the above equation system, each term of the form $\varnothing(M \Rightarrow N)$ stands for a single variable. Solving the equation system Eqn. (4.12) yields, among other things, that the average reduction length of $M_2 \Rightarrow n_i$ equals six. To see this, again transform the above equation system by upwards variable substitution into the following set of equations:

$$\varnothing(M_2 \Rightarrow n_i) = 1 + 1 + 1 + 0.5 \cdot \varnothing(M_2 \Rightarrow n_i) \tag{4.13}$$

$$\varnothing((\lambda x.(x|n_i))M_2 \Rightarrow n_i) = 1 + 1 + 0.5 \cdot \varnothing(M_2 \Rightarrow n_i) \tag{4.14}$$

$$\varnothing(n_i|M_2 \Rightarrow n_i) = 1 + 0.5 \cdot \varnothing(M_2 \Rightarrow n_i) \tag{4.15}$$

$$\varnothing(n_i \Rightarrow n_i) = 0 \tag{4.16}$$

Now, solving Eqn. (4.13) yields six for $\varnothing(M_2 \Rightarrow n_i)$, i.e., for the average number of steps of $M_2 \Rightarrow n_i$.

All that has been said for the program M_2 is also valid for the program M_3 with the uniquely existing non-terminating program run ω_3, which can be illustrated as follows:

$$\begin{aligned}\omega_{M_3} &= \mu\lambda x.(-1(+1(x)) \mid n_i) \xrightarrow{1} \cdots \xrightarrow{0.5} \\ &-1(+1(M_3)) \xrightarrow{1} \cdots \xrightarrow{0.5} \\ &-1(+1(-1(+1(M_3)))) \xrightarrow{1} \cdots\end{aligned}$$

All the arguments that we have used with respect to M_2 immediately apply to the program M_3. Nevertheless, there is a very important difference between the programs M_2 and M_3. The execution of M_2 yields at most a finite number of terms, even in case of the non-terminating program run ω_2. This is different in the case of M_3. Here, the number of terms reachable by program execution is uncapped, i.e., an infinite number of terms can be reached, which encompass all terms of the following forms for all numbers n:

$$\underbrace{(-1(+1(}_{n\ times}\ldots(n_i)\ldots))) \qquad \underbrace{(-1(+1(}_{n\ times}\ldots M_3\ldots)))$$

This is an important observation. Let us call the set of terms reachable by execution of a program the term cover of this program, or cover of the program for short. We can exploit the fact that the cover of M_2 is finite to determine its vectors of reduction probabilities and average execution lengths, algorithmically, via solution of a finite equation system. This is not the case for M_3, because the cover of M_3 is infinite. Fortunately, finite term covers are computable. This means that there exists an algorithm that terminates and correctly computes the term cover for a program, whenever this term cover is finite. This will be the topic of the upcoming sections. Actually, the finiteness of term covers characterizes a widened class of termination behavior that we will call path stoppability, or p-stoppability, later.

4.1.1 The Complete Picture of Termination Behavior

Let us complete the big picture of termination behavior with all of its aspects that are discussed in this chapter. As shown in Table. 4.3 we deal with four different aspects. These aspects are the termination degree, the existence of a non-terminating program run, the boundedness of a program and the finiteness of the cover of a program. We consider the case that the termination degree equals one and the case that the termination degree is less than one. As in Table 4.1, the existence of a non-terminating program run is indicated by $T(M) \subset P(M)$. Next, $\beta(M)$ indicates that a program is bounded, whereas $\overline{\beta}(M)$ indicates that a program M is unbounded; compare with Defs. 4.3 and 4.4. The infinity sign ∞ in Table. 4.3 indicates that the cover of a program $Cover(M)$ is infinite, whereas $\overline{\infty}$ indicates that the cover is finite.

	$\eta_S\langle M, C\rangle = 1$		$\eta_S\langle M, C\rangle < 1$		
$T(M)$	M_1	$\times_{(i)}$	$\times_{(ii)}$	$\times_{(iii)}$	$\overline{\infty}$
$=$ $P(M)$	$\times_{(iv)}$	$\times_{(v)}$	$\times_{(vi)}$	$\times_{(vii)}$	∞
$T(M)$	$\times_{(viii)}$	M_2	M_0	M_5	$\overline{\infty}$
$\subset$ $P(M)$	$\times_{(ix)}$	M_3	M_6	M_7	∞
	$\beta(M)$	$\overline{\beta}(M)$	$\beta(M)$	$\overline{\beta}(M)$	

Table 4.3. Different aspects of termination behavior

In principle, given the four dimension with two possible instances each, we have sixteen possible combinations in Table 4.3. Only some part of these combinations are possible. The impossible combinations are indicated by a cross in Table 4.3. It is the task of this chapter to prove these combinations to be impossible. Now, for each possible combination we have given an example in Table 4.3. All the example programs that have been discussed earlier in this section occur in Table 4.3. The discussed programs M_1, M_2 and M_3 can all be found in the column of termination degree one. Actually, with respect to a termination degree of one these three examples are exhaustive. All other combinations of termination degree one are impossible. First, in case of termination degree one, the boundedness of a program is determined by the existence of a non-terminating program run. A program is unbounded if and only if it has a non-terminating program run. Then, a program that has an infinite cover necessarily has a non-terminating program run.

A program run that has a termination degree less than one must have a non-terminating program run. Therefore, there remain four combinations for which we have not yet discussed existing examples. These programs can be given, e.g., by $M_0 = (n_i \mid \mu\lambda x.x)$, $M_5 = (M_2 \mid \mu\lambda x.x)$, $M_6 = (n_i \mid \mu\lambda x.{+1}(x))$, and $M_7 = (M_3 \mid \mu\lambda x.x)$.

4.1.2 Outline of the Propositions on Termination Behavior

Let us give an overview of the propositions proved in this chapter and see how they relate to the overall picture provided by Table 4.3.

- *Eqn. (4.40)* The termination degree of a program equals the probability of its terminating program runs.
- *Lemma 4.32* If a program has an infinite cover then it has a non-terminating program run.
- *Lemma 4.39* If a program terminates unbounded then it has a non-terminating program run.

- *Lemma 4.40* If a program has termination degree of one and has a non-terminating program run then it terminates unbounded.

Together, Eqn. (4.40) and Lemmas 4.32, 4.39 and 4.40 prove all the fields in (i) through (ix) in Table 4.3 impossible. Based on the definitions of termination degree in Def. 4.1 and terminating program runs in Def 4.18, Eqn. (4.40) proves fields (ii), (iii), (vi) and (vii) impossible. Lemma 4.32 proves fields (iv), (v), (vi) and (vii) impossible. Lemma 4.39 proves fields (i), (v), (iii) and (vii) impossible. Lemma 4.40 proves fields (i), (v), (viii) and (ix) impossible.

Although there is some redundancy between the above propositions, i.e., overlap in proving fields in Table 4.3 impossible, all of the propositions are necessary. Eqn. (4.40) is essential to prove field (ii) impossible, Lemma 4.32 is essential to prove field (iv) impossible, Lemma 4.39 is essential to prove field (i) impossible, and Lemma 4.40 is essential to prove fields (viii) and (ix) impossible.

4.2 Bounded and Unbounded Termination

With the notion of bounded termination we model that the termination of a program never exceeds a known maximum number of steps. A program terminates bounded if all of its terminating program runs terminate in at most a known maximum number of steps. Otherwise, we say that the program terminates unbounded.

We now define bounded and unbounded termination in terms of probabilities, or, to be precise, in terms of hitting probabilities and bounded hitting probabilities. Given a program M, we say that it terminates bounded if its degree of termination is already determined after a maximum number of finitely many steps. This means that its bounded hitting probability to reach a constant does not increase any more after a certain maximum number of steps.

Definition 4.3 (Bounded Termination) *Given a program M we say that it terminates bounded, denoted by $\beta(M)$,* **iff** *there exists a bound $n \in \mathbb{N}_0$ such that the program's probability to reach a constant after at most n steps equals its termination degree, i.e.,*

$$\eta_S^n\langle M, C\rangle = \eta_S\langle M, C\rangle \tag{4.17}$$

Now, unbounded termination is defined as the dual notion of bounded termination, i.e., a program is considered to be terminating unbounded whenever it does not terminate bounded.

Definition 4.4 (Unbounded Termination) *Given a program M we say that it terminates unbounded, denoted by $\overline{\beta}(M)$,* **iff** *the program does not terminate bounded, i.e., $\neg\beta(M)$, i.e.,*

$$\forall n \in \mathbb{N}_0 \;.\; \eta_S^n\langle M, C\rangle \neq \eta_S\langle M, C\rangle \tag{4.18}$$

Next, Lemma 4.5 once more illustrates why we are speaking about bounded resp. unbounded termination. Of course, for all programs the degree of termination is limited, absolutely, because no program can have a termination degree greater than one. However, in case of unbounded termination, the degree of termination never stops increasing as expressed by Lemma 4.5.

Lemma 4.5 (Monotonicity of Bounded Termination Degrees)
Given a program M that terminates unbounded, i.e., $\overline{\beta}(M)$, we have that:

$$\forall n \,.\, \exists i > n \,.\, \eta^{i}_{S}\langle M, C\rangle > \eta^{n}_{S}\langle M, C\rangle \tag{4.19}$$

Proof. The Lemma is proven by contraposition. Assume a program M such that there exists an n such that for all $i > n$ we have that $\eta^{i}_{S}\langle M, C\rangle \leqslant \eta^{n}_{S}\langle M, C\rangle$. Actually, we know that $\eta^{i}_{S}\langle M, C\rangle = \eta^{n}_{S}\langle M, C\rangle$ for all $i > n$, because by Lemma 2.28 we know that $\eta^{i}_{S}\langle M, C\rangle$ in monotonically increasing and therefore it is impossible that $\eta^{i}_{S}\langle M, C\rangle < \eta^{n}_{S}\langle M, C\rangle$. Furthermore, due to Corollary 2.27 we know that $\lim_{i \to \infty} \eta^{i}_{S}\langle M, C\rangle = \eta_{X}\langle M, C\rangle$ and therefore also $\eta^{n}_{S}\langle M, C\rangle = \eta_{X}\langle M, C\rangle$, which means that M is bounded. □

Let us call the probability that a program reaches a constant after at most n steps its bounded termination degree after n steps, or termination degree after n steps for short. Lemma 4.5 expresses that the bounded termination degree is more than only monotonically increasing. It is not strictly increasing, yet it is more than only monotonically increasing. Independent of the number of steps already taken, we can be sure that the termination degree will eventually increase further. The fact that the bounded termination degree is monotonically increasing is not the point. This already follows immediately from the fact that each bounded hitting probability is monotonically increasing, see Lemma 2.28, and is, moreover, independent of the question whether a program is bounded or not.

The question whether a program terminates bounded or unbounded is independent of its termination degree. In particular, we also speak about bounded and unbounded program termination in case of programs with termination degree less than one, as indicated by Table 4.1. For, example, the program $0|\mu\lambda x.x$, which has a termination degree of 0.5, terminates bounded, because it has only one terminating program run. Actually, even in case of the program $\mu\lambda x.x$, which has no terminating program runs at all, we say that the program terminates bounded. Bounded termination means only that all the terminating program runs of a program have a maximum execution length, it does not say anything about the termination degree.

Irrespective of all this, we are particularly interested in discussing the boundedness of program termination in case of programs that have a termination degree of one. Again, as indicated by Table 4.1, in case of a termination degree of one, the question whether a termination is bounded or unbounded is determined by whether all program runs of a program terminate or there exists a non-terminating program run. We have observed this already in our examples, however, we have not yet proven this. Although this might be intuitively clear, we have not yet developed the appropriate technical apparatus to precisely prove this fact. We need to defer the proof of this fact to Sect. 4.8 after we have defined reduction graphs in Sect. 4.4 and reduction trees in Sect. 4.7.

4.3 Program Executions and Program Runs

4.3.1 Program Executions

Basically, program executions are outcomes of the Markov chains semantics. However, we do not want to consider those outcomes as program executions that contain a step or steps with a zero percent probability. In general, infinitely many such outcomes exist for each program, however, they are not intended as program executions, i.e., they are unwanted as program executions. In the sequel we will refer to such outcomes as spurious outcomes. This choice of terminology is an arbitrary choice and maybe not even the best one. Other options would be unwanted, undefined or false outcomes.

A program execution of a program M is a process instance starting from M in which all steps have a non-zero percent probability. We denote the set of program executions of a given program M by $P_\varepsilon(M)$.

Definition 4.6 (Program Executions) *Given a program M, the set $P_\varepsilon(M) \subseteq \Omega_S$ of its program executions is defined as follows:*

$$P_\varepsilon(M) = \{\ \omega \in (S_0 = M) \mid \forall i \in \mathbb{N}_0 \ .\ S_i(\omega) \to S_{i+1}(\omega) > 0\ \}$$

With respect to the above definition, it might be helpful to recall some of the Markov chain syntax. Given a program M, the set of process instances starting from M is given by the set $S_0 = M$. Remember, that $(S_0 = M) \subseteq \Omega_S$ is the Markov chain notation to denote $S_0^{-1}(M)$, i.e., the inverse image of M under S_0 – see Def. 2.10. It is also instructive to make explicit the single process instances in the set $S_0 = M$ as in $\{\ \omega \in \Omega_S \mid S_0(\omega) = M\ \}$.

The intention of Def. 4.6 is to restrict the set of program executions to those process instances that have no steps with a zero percent probability. The required probability is expressed in Def. 4.6 in terms of the probability matrix. Recall, that this is entirely adequate, because due to Corollary 2.15 we know that for all $i \in \mathbb{N}_0$ we have that

$$\mathsf{P}(S_{i+1} = S_{i+1}(\omega) \mid S_i = S_i(\omega)) = S_i(\omega) \to S_{i+1}(\omega)$$

Next, we define the set of spurious outcomes of a program M, which we denote by $\overline{P_\varepsilon}(M)$. Remember that a spurious outcome is an outcome that is not a program execution. Consequently, given a program M, its spurious outcomes are defined as the complement of $P_\varepsilon(M)$ in $S_0 = M$.

Definition 4.7 (Spurious Outcomes) *Given a program M, the set $\overline{P}_\varepsilon(M)$ of its spurious outcomes is defined as follows:*

$$\overline{P}_\varepsilon(M) = (S_0 = M) \backslash P_\varepsilon(M) \tag{4.20}$$

A spurious outcome has a zero percent probability, because it contains at least one transition with a zero percent probability. This is intuitively clear and can be proven easily by natural induction. Sets of spurious outcomes all have a zero percent probability. Similarly, the probability that a program is executed as one of its program executions is a hundred percent. Although these facts are intuitively clear, they are non-trivial. Lemma 4.8, Corollary 4.9 and Lemma 4.10 show such and similar facts in the sequel.

Lemma 4.8 (Probability of Program Executions) *Given a program M, the probability of its set of program executions equals its initial probability, i.e., we have that*

$$\mathsf{P}(P_\varepsilon(M)) = \iota_M \tag{4.21}$$

Proof. We define the n-approximating set to $P_\varepsilon(M)$, denoted by $P_{<n}(M)$, as follows:

$$P_{<n}(M) = \{\ \omega \in (S_0 = M) \mid \forall i < n \ .\ S_i(\omega) \rightarrow S_{i+1}(\omega) > 0\ \} \tag{4.22}$$

It can be shown by natural induction that the set $S_n^\dagger(P_{<n}(M))$ is finite for all n. Similarly, it can be shown by natural induction that $\mathsf{P}(P_{<n}(M))$ equals ι_M for all n. Now we know that $P_{<n}(M) \supseteq P_{<n+1}(M)$ for all n and therefore that $P_{<0}(M) \supseteq P_{<1}(M) \supseteq \cdots$ forms a decreasing chain. Therefore, by continuity from above, i.e., Lemma 2.8, we know that

$$\mathsf{P}(\bigcap_{n \in \mathbb{N}_0} P_{<n}(M)) = \lim_{n \rightarrow \infty} \mathsf{P}(P_{<n}(M)) = \lim_{n \rightarrow \infty} \iota_M = \iota_M \tag{4.23}$$

Furthermore, by appropriate set transformations and application of De Morgan transformations it can be shown that

$$P_\varepsilon(M) = \bigcap_{n \in \mathbb{N}_0} P_{<n}(M) \tag{4.24}$$

Now, Eqn. (4.21) follows immediately from Eqns. (4.23) and (4.24). □

Given a program, the event that one of its program executions is executed upon program start has a hundred percent probability, whereas the event that one of its spurious outcomes is executed has a zero percent probability. These facts are expressed by Corollary 4.9.

Corollary 4.9 (Probability of Program Executions) *Given a program M we have that*

$$\mathsf{P}_M(P_\varepsilon(M)) = 1 \tag{4.25}$$

$$\mathsf{P}_M(\overline{P}_\varepsilon(M)) = 0 \tag{4.26}$$

Proof. Corollary from Lemma 4.8. □

Spurious outcomes do not add to probabilities at all. They can be erased from any event without an effect on its probability. This is expressed by the next Lemma 4.10.

Lemma 4.10 (Restriction to Program Executions) *Given a program M and an event $A \subseteq \Omega_S$ we can restrict A to its subset of program executions without changing its probability conditional on M, i.e.,*

$$\mathsf{P}_M(A) = \mathsf{P}_M(A \cap P_\varepsilon(M)) \tag{4.27}$$

Proof. First, we show the following:

$$\mathsf{P}(A \cap S_0 = M) = \mathsf{P}(A \cap P_\varepsilon(M)) \tag{4.28}$$

By the definition of spurious outcomes in Def. 4.7 we know that $S_0 = M$ equals $P_\varepsilon(M) \cup \overline{P}_\varepsilon(M)$. Furthermore, we know that $P_\varepsilon(M)$ and $\overline{P}_\varepsilon(M)$ are disjoint. Therefore we know that

$$\mathsf{P}(A \cap S_0 = M) = \mathsf{P}(A \cap P_\varepsilon(M)) + \mathsf{P}(A \cap \overline{P}_\varepsilon(M)) \tag{4.29}$$

By Corollary 4.9 we know that $\mathsf{P}_M(\overline{P}_\varepsilon(M)) = 0$. Therefore, we also know that $\mathsf{P}(\overline{P}_\varepsilon(M)) = 0$ and also know that $\mathsf{P}(A \cap \overline{P}_\varepsilon(M) = 0$. Taking this together with Eqn. (4.29), we now know that $\mathsf{P}(A \cap S_0 = M)$ equals $\mathsf{P}(A \cap P_\varepsilon(M))$, i.e., we know that Eqn. (4.28) holds. Now, Eqn. (4.27) immediately follows as a corollary. We know that $\mathsf{P}_M(A)$ equals $\mathsf{P}(A \cap S_0 = M)/\mathsf{P}(S_0 = M)$ which, according to Eqn. (4.28), equals $\mathsf{P}(A \cap P_\varepsilon(M) \cap S_0 = M)/\mathsf{P}(S_0 = M)$, which equals $\mathsf{P}_M(A \cap P_\varepsilon(M))$. □

Next, we define the set of terminating program executions starting from a program M, which we denote by $T_\varepsilon(M)$. A terminating program run is a program run that eventually hits a constant. Dually, a non-terminating program run is a program run that never hits a constant and we denote the set of non-terminating programs by $\overline{T}_\varepsilon(M)$.

Definition 4.11 (Terminating Program Executions) *Given a program M, the set of its terminating program executions, denoted by $T_\varepsilon(M) \subset \Omega_S$, and the set of its non-terminating program executions, denoted by $\overline{T}_\varepsilon(M)$, are defined as follows:*

$$T_\varepsilon(M) = (\exists\, i<\infty \,.\, S_i \in C) \cap P_\varepsilon(M)$$
$$\overline{T}_\varepsilon(M) = P_\varepsilon(M) \backslash T_\varepsilon(M)$$

According to Corollary 2.25, $T_\varepsilon(M)$ can also be characterized differently as $(H_X(T) \leqslant n \,\cap\, P_\varepsilon(M))$. Now, due to Lemma 4.10 we have that each event can be restricted to its program executions without changing its probabilities. Therefore, and due to the definitions of degrees of termination and hitting probabilities, see Defs. 4.6 and 2.23, we have that a program's degree of termination equals the probability of this program's terminating executions, which is again intuitively immediately clear:

$$\eta_S\langle M, C\rangle = \mathsf{P}_M(T_\varepsilon(M)) \tag{4.30}$$

Similarly, we have that a program's degree of non-termination equals the probability of this program's non-terminating executions:

$$\eta_S\langle M, \overline{C}\rangle = \mathsf{P}_M(\overline{T}_\varepsilon(M)) \tag{4.31}$$

For technical reasons of Markov chains we need to distinguish between program executions and program runs. A program execution is an outcome of the Markov chain semantics. A program run is a class of program executions that cannot be distinguished any further by the Markov chain semantics. This distinction should not be overstressed in argumentations, as least not at the informal or conceptual level. Nevertheless, it is an important distinction that must be maintained throughout the development of the subsequent theory. In particular, we will need to face this distinction in proofs. Let us stop and delve into this topic for a while in Sect. 4.3.2.

4.3.2 Program Runs in the Sigma Algebra

A program run is a specification that fixes a program for each stage of our Markov chain semantics, i.e., it is an infinite sequence $(s_i \in \Lambda_P)_{i\in\mathbb{N}_0}$. In due course, in Sect. 4.4, we will define a concrete concept of program runs as infinite walks in a reduction graph. Now, given such a program run $s = (s_i \in \Lambda_P)_{i\in\mathbb{N}_0}$, in general, we cannot assume that the program execution that adheres to s is uniquely defined. This means that it is not guaranteed, that the set $\{\omega \mid \forall i.S_i(\omega) = s_i\}$ consists of exactly one element. Note that a single program execution ω has – technically – no probability, i.e., only taken as an event $\{\omega\}$ it receives a probability $\mathsf{P}(\{\omega\})$. Now, let us say that two program executions ω_1 and ω_2 are equal, denoted by $\omega_1 \equiv_S \omega_2$, if they cannot be distinguished by S, i.e., $S_i(\omega_1) = S_i(\omega_2)$ for all $i \in \mathbb{N}_0$.

Now, we say that a program run is an equivalence class of program executions under $\equiv_S$. Note that in Def. 4.12 we use quotient set notation as introduced in Def. 2.87 in order to denote the set $\Omega_{S/\equiv_S}$ of all relevant equivalence classes of program executions.

Definition 4.12 (Program Run Events in the Sigma Algebra) *An event $p \in \Sigma_S$ is called a program run event, or just program run for short, if $p \in \Omega_{S/\equiv_S}$, where $\equiv_S$: $\Omega_S \times \Omega_S$ is defined as image equality under S, i.e., given ω_1 and ω_2, we have that $\omega_1 \equiv_S \omega_2$* **iff** *$S_i(\omega_1) = S_i(\omega_2)$ for all $i \in \mathbb{N}_0$.*

Things would change and all of the discussed technical issues would perhaps become easier, if we required the random variables in S to be injective, however, we prefer to stay with the general case. Conceptually, we can forget about the distinction between program executions and program runs in upcoming argumentations. This is so, because in the sequel we are never interested in the probabilities of distinct, equivalent program executions ω_1 and ω_2 but always only in the probability of their joint equivalence class. However, the situation would change, if we introduced, besides the random variables in S, further auxiliary random variables of the form $X : \Omega_S \rightarrow S'$ that model different aspects of resp. more fine-grained perspectives on the program executions. However, we are only interested in S in the sequel, with which, by its definition, ω_1 and ω_2 cannot be distinguished.

In the sequel we freely use the term program run for equivalence classes of program executions, unique specifications of infinite program sequences, and concrete concepts for specifications of program sequences such as the reduction walks in our reduction path. In informal descriptions we often neglect the difference between a program run and its single executions. Then, we state a property about the program run which is actually an all-quantified property about its program executions. For example, we might say that a program run is terminating, which actually means that all of its program executions terminate. Or we say that a program run eventually reaches a certain state. Again we mean that all of its program executions reach that state. All of this is acceptable, at least in informal descriptions, given that the single program executions of a program run cannot be distinguished merely by the Markov chain semantics itself.

4.4 The Reduction Graph

The one-step semantics $M \xrightarrow{i} N$ expresses the possibility of an immediate reduction from closed term M to closed term N with probability i. For the analysis of termination behavior we need to investigate the finite reachability of terms, to which we can abstract from the probabilities of the choices during program executions. We are interested in different kinds of reducibility relations, reduction paths and reducibility predicates that equip us with the necessary properties of term reductions. All of these notions can be defined on the basis of graphs and walks in graphs.

As a first step we define the one-step reduction relation, also called the immediate reduction relation, on the basis of the one-step semantics. The immediate reduction relation forgets about probabilities in the one-step semantics and, hand-in-hand with this, erases steps that have a zero probability. The intention behind this is clear. We want to single out spurious program executions. The immediate reduction relation is denoted by ρ. It is therefore defined as a binary relation over the set of programs Λ_P.

Definition 4.13 (The Immediate Reduction Relation) *The immediate reduction relation of the probabilistic lambda calculus, denoted by ρ, is defined* *as follows:*

$$\rho \subseteq \Lambda_P \times \Lambda_P \tag{4.32}$$
$$\rho = \{\langle M, N\rangle \mid M \to N > 0\} \tag{4.33}$$

Next, we define the reduction graph R of the lambda calculus as the graph consisting of the set of programs Λ_P as nodes and the reduction relationships defined by ρ as edges.

Definition 4.14 (The Reduction Graph) *The reduction graph of the probabilistic lambda calculus, denoted by R, is defined as a digraph $\langle V_R, E_R\rangle$ with nodes V_R and edges E_R as follows:*

$$R = \langle V_R, E_R\rangle \tag{4.34}$$
$$V_R = \Lambda_P \tag{4.35}$$
$$E_R = \rho \tag{4.36}$$

Now, reduction walks are walks in the reduction graph. We call reduction walks also multi-step reductions or simply reductions for short. Similarly, we call the paths in the reduction graph reduction paths. We inherit all definitions for walks and paths in graphs. This way we have the set nW_R of finite reductions of length n, compare with Eqn. (2.53), the set $\star W_R$ of finite reductions of arbitrary length, compare with Eqn. (2.54), the set ωW_R of infinite reductions, compare with Eqn. (2.55), and the set $\oplus W_R$ of all reductions, compare with Eqn. (2.56). Given a set of walks $W \subseteq \oplus W_R$ we denote the subset of its reductions starting with a term M by $W(M)$, compare with Def. 2.37. Reduction paths are those reductions in which no term occurs more than once and, given a set of reduction walks $W \subseteq \oplus W_R$, we denote its subset of reduction paths by $\pi(W)$; compare with Def. 2.38. Equally, all other notation available for graphs applies immediately to the reduction graph and its parts; see, e.g., Def. 2.36. Recall that, given a graph G and a walk w, we use $|G|$ and $|w|$ to denote their sizes, $\kappa(G)$ and $\kappa(w)$ to denote the sets of their nodes, and $\#(w)$ to denote the length of a walk.

4.4.1 Program Runs in the Reduction Graph

Now, the infinite walks of the program reduction graph R will serve as a model for program runs. An infinite walk in the reduction graph fixes a state for each stage of the program execution. An infinite walk in the reduction graph uniquely identifies a set of program executions that cannot be distinguished further by the Markov chain S, i.e., a program run of our Markov chain semantics; see Def. 4.12. Henceforth, we will naturally identify infinite graph walks with events of our Markov chain semantics. Henceforth, we will also talk about infinite graph walks simply as program runs.

Definition 4.15 (Program Runs in the Reduction Graph) *A program run p is an infinite walk of the reduction graph R, i.e., $p \in \omega W_R$. Given a program M, the set $P(M)$ of its program runs is defined as the set of its infinite walks starting from M, i.e.,*

$$P(M) = \omega W_R(M)$$

We use only the infinite walks of the reduction graph as models of the process instances of our Markov chain semantics. Nonetheless, the whole set $\oplus W_S$ of all walks in the reduction graph is important for us. We need finite walks as approximations to program runs in informal arguments as well as in formal proofs of program properties. We might call the finite walks in the reduction graph $\star W_S(M)$ the approximating program runs of program M. Things will be different later in Sect. 4.7 when we introduce the notion of reduction tree. We will use the infinite paths of the reduction tree to model the non-terminating program runs and nothing but the non-terminating program runs. In the reduction tree, also some of the finite paths are considered as program runs, i.e., they are used to model the terminating program runs. The reduction graph and the reduction tree serve different purposes. We use the reduction graph to analyze a widened class of termination, the so-called path stoppability. We use the notion of reduction tree to analyze boundedness of program termination. The reduction graph comes first. This means that reduction trees are defined on the basis of the reduction graph; actually, each reduction tree is defined in terms of the walks in the reduction graph.

Program runs are formalized as sequences of terms and not as sequences of alternating terms and probabilities. The reduction graph is a digraph, not a multigraph, and all digraph walks are already completely determined by their nodes – see also the remarks on walks in digraphs that follow Def. 2.35. This is appropriate, because, as we know from Lemma 3.3, the one-step semantics is a function with respect to pairs of terms. Given a program run $r_0 r_1 r_2 \ldots$ we immediately know its form as $r_0 \xrightarrow{i_0} r_1 \xrightarrow{i_1} r_2 \xrightarrow{i_2} \ldots$. Although the first form is the form that will be with us in proofs, the second one is the one that we usually prefer in informal discussions, i.e., in examples we will depict a program run also in that second form.

A program run stands for a set of program executions. A program run fixes a state for each stage of a program execution. A program execution ω adheres to a program run r if it is in state r_i for all stages $i \in \mathbb{N}_0$, i.e., whenever $S_i(\omega) = r_i$. Then, we say that the program execution ω is a program execution of the program run r and, vice versa, that r is the program run of program execution ω. We denote the program run of a program execution ω by $\varrho(\omega)$ and the set of program executions of a program run r by $\varepsilon(r)$. Obviously, all program executions of a program run are equal with respect to S, i.e., they cannot be distinguished only in terms of S. Actually, a set $\varepsilon(r)$ is a program run event in the sense of Def. 4.12, i.e., $\varepsilon(r) \in \Omega_{S_{/\equiv_S}}$.

Definition 4.16 (Program Run of a Program Execution) *Given a program execution $\omega \in P_\varepsilon(M)$ we define its program run, denoted by $\varrho(\omega)$, as follows:*

$$\varrho(\omega) = (S_i(\omega))_{i \in \mathbb{N}_0}$$

Definition 4.17 (Executions of a Program Run) *Given a program run $r \in P(M)$ and a set of program runs $R \subseteq P(M)$ we define the set of program executions of r, denoted by $\varepsilon(r)$, and the set of program executions of R, denoted by $\varepsilon(R)$, as follows:*

$$\varepsilon(r) = \varrho^{-1}(r)$$
$$\varepsilon(R) = \cup\, \varepsilon^{\dagger}(R)$$

Obviously, given a program M, we have that the set of its program executions as defined by Def. 4.6 equals the set of program executions of all of its program runs, i.e.,

$$P_\varepsilon(M) = \varepsilon(P(M)) \tag{4.37}$$

At least in informal descriptions and argumentations, we can speak of a program run as if it were an event. In particular, we can say that a program run has a probability. We can do so, exactly because of the one-to-one correspondence between the program run events and graph walks as described above. Formally, the probability of a program run $r = w_0 w_1 w_2 \cdots$ can be immediately defined as the probability $\mathsf{P}_{w_0}(\varepsilon(r))$. These extra technical and terminological efforts are complicated only at a first sight. In concrete argumentations the distinctions between the two kinds of program runs do not play a major role. On the other hand, the efforts pay back a lot. We have set the stage for graph theory to argue about program runs and program executions. Despite the fact that the clear distinction between the event facet and the graph facet of a program run can be neglected in informal discussions, we stay with the dualism when we give further definitions.

4.4.2 Terminating and Non-Terminating Program Runs

Next, we turn to the notions of terminating and non-terminating program runs. A terminating program run is a program run that eventually reaches a constant. Dually, a non-terminating program run is a program run that never reaches a constant. Given a program M, we denote its terminating program runs by $T(M)$, whereas we denote its non-terminating program runs as $\overline{T}(M)$.

Definition 4.18 (Terminating Program Runs) *Given a program M, the set $T(M) \subseteq \omega W_S$ of its terminating program runs and the set $\overline{T}(M) \subseteq \omega W_S$ of its non-terminating program runs are defined as follows:*

$$T(M) = \{ w \in P(M) \mid \exists i < \infty \,.\, w_i \in C \} \tag{4.38}$$
$$\overline{T}(M) = P(M) \backslash T(M) \tag{4.39}$$

In Eqn. (4.37) we have seen how $\varepsilon : \mathbb{P}(\omega W_S) \to \mathbb{P}(\Omega_S)$ equates program runs to program executions, i.e., $P_\varepsilon(M) = \varepsilon(P(M))$. Similar correspondences exist for terminating program runs and non-terminating program runs, i.e., we have that $T_\varepsilon(M) = \varepsilon(T(M))$ and $\overline{T}_\varepsilon(M) = \varepsilon(\overline{T}(M))$. In particular, with respect to degrees of termination and non-termination we can broaden Eqns. (4.30) and (4.31) to program runs and summarize them as follows:

$$\begin{aligned} \eta_S\langle M, C\rangle &= \mathsf{P}_M(T_\varepsilon(M)) = \mathsf{P}_M(\varepsilon(T(M))) \\ \eta_S\langle M, \overline{C}\rangle &= \mathsf{P}_M(\overline{T}_\varepsilon(M)) = \mathsf{P}_M(\varepsilon(\overline{T}(M))) \end{aligned} \tag{4.40}$$

In the sequel, we will use any of the characterizations shown in Eqn. (4.40) interchangeably to denote degrees of termination resp. degrees of non-termination without further comment.

An important observation is that a terminating program run cannot be a path in R. This is so, because a terminating program run contains at least the circle $\ldots c \overset{1}{\to} c \ldots$ for the constant c that it eventually reaches. To say it differently, as the contrapositive, each program run that is a path is automatically non-terminating. The converse is not true. There exist many program runs that are non-terminating but are not automatically paths. Take, as an example, the most simple non-terminating program $\mu\lambda x.x$. This program executes as the program run $\mu\lambda x.x \overset{1}{\to} (\lambda x.x)\mu\lambda x.x \overset{1}{\to} \mu\lambda x.x \overset{1}{\to} \ldots$, which is obviously not a path.

The fact that a terminating program run cannot be a path is sufficient to prove that programs that have an infinite cover necessarily have a non-terminating program run; see Lemma 4.32. Actually, Lemma 4.32 follows more or less immediately from this fact against the background of König's Lemma; see Lemma 2.39. However, the fact is not sufficient to prove that every unbounded program necessarily has a non-terminating program run; see Lemma 4.39. In order to prove Lemma 4.39 we need some stronger model of non-terminating program run later. With the reduction tree in Sect. 4.7 we turn each non-terminating program run into an infinite tree path, whereas terminating program runs become finite tree paths. With this correspondence it will be then be possible to prove Lemma 4.39.

The fact that a terminating program run cannot be a path follows from the fact that terminating programs stabilize behind the constant they reach. A further property that we need in proofs later is the fact that the constant that is reached by a terminating program run is uniquely determined. We summarize these facts in the following in Lemmas 4.19 and 4.20 and Corollary 4.21.

Lemma 4.19 (Terminating Program Runs Stabilize) *Given a program M, each of its terminating program runs $p \in T(M)$ stabilizes behind the constant it reaches, i.e., for all $p \in T(M)$, $i \in \mathbb{N}_0$ and $j > i$ such that $p_i \in C$ we have that $p_j = p_i$.*

Proof. Let us assume that $p_i = c$ for some constant $c \in C$ and $i \in \mathbb{N}_0$. We proceed by natural induction over the sequence index $j \geqslant i$. The case $j = i$ is trivial with $p_i = c$. In case $j \geqslant 0$ we can assume that $p_j = c$. Due to the definition of the one-step semantics, rules (3.62) and (3.63), we know that $c \overset{1}{\to} c$ is the only transition with $c \overset{1}{\to} N > 0$. Therefore we know due to the definition of $P(M)$ in Def. 4.15 that p_{j+1} must equal $c = p_j$. □

Lemma 4.20 (Program Runs Terminate Uniquely) *Given a program M, the constant that is reached by one of its terminating program runs $p \in T(M)$ is uniquely given, i.e., for all $p \in T(M)$ there exists a $c \in C$ such that for all $i \in \mathbb{N}_0$ we have that $p_i \in C$ implies $p_i = c$.*

Proof. Due to Def. 4.18 we know that there exists a $c \in C$ such that $p_k = c$ for some $k \in \mathbb{N}_0$. Now assume that $p_l \in C$ for some $l \in \mathbb{N}_0$. Now for all $l \in \mathbb{N}_0$ either $l = k$, $k > l$ or $l > k$. The case $i = k$ is trivial. In case $k > l$ we know due to Lemma 4.19 that $c = p_k = p_l$. In case $l > k$ we know due to Lemma 4.19 that $p_l = p_k = c$. □

Corollary 4.21 (Terminating Program Runs are not Paths) *If a program run $p \in P(M)$ is terminating, then p is not a path, i.e., we have that $p \in T(M)$ implies $p \notin \pi(P(M))$.*

Proof. Corollary to Lemma 4.19. □

Corollary 4.21 amounts to saying that the set of terminating program runs and the set of paths are disjoint, i.e., $T(M) \cap \pi(P(M)) = \emptyset$.

Now, we turn to the central concepts of termination behavior that we have introduced in Defs. 4.1 and 4.3, i.e., termination degree and boundedness of termination. We want to see how these notions meet the intuition behind the program runs that we have just introduced. We turn to the notion of termination degree first. The termination degree of a program M has been defined as the hitting probability that our Markov chain reaches a constant by starting from M, i.e., informally, the probability that M reaches a constant. In terms of program runs, the termination degree will turn out to be the joint probability of all terminating program runs of M, i.e., the probability that M executes as one of its terminating program runs. The latter is again, informally, the probability that M reaches a constant.

Next, let us turn to the notion of bounded termination and its relationship to the notion of program runs as defined for the reduction graph. We have defined bounded termination as the property that the termination degree of a program is already known after a fixed number of steps. In terms of program runs a program is bounded whenever the number of steps needed to reach a constant is capped. This is expressed by Lemma 4.22.

Lemma 4.22 (Bounded Termination) *A program M terminates bounded* **iff** *there exists a bound $n < \infty$ so that for all terminating program runs $t \in T(M)$ there exists an $i \leqslant n$ such that $t_i \in C$, i.e.,*

$$\beta(M) \Leftrightarrow \exists n{<}\infty \,.\, \forall t \in T(M) \,.\, \exists\, i \leqslant n \,.\, t_i \in C \tag{4.41}$$

Proof. We start with the characterization of bounded termination in terms of program runs, i.e., the right-hand side of the equivalence in Eqn. (4.41). Given that $T(M)$ equals $\{t \mid \exists i{<}\infty.t_i{\in}C,\ t \in P(M)\}$ we can write the right-hand side of Eqn. (4.41) differently in the following form:

$$\exists n{<}\infty \,.\, \{t \mid \exists i{<}n.t_i{\in}C,\ t \in P(M)\} \supseteq \{t \mid \exists i{<}\infty.t_i{\in}C,\ t \in P(M)\} \tag{4.42}$$

In the course of the proof, we will henceforth use $T[n]$ to denote the set $\{t \mid \exists i < n.t_i \in C,\ t \in P(M)\}$ from Eqn. (4.42) and T' to denote the set $\{t \mid \exists i < \infty.t_i \in C,\ t \in P(M)\}$. Now, the subset relationship $T[n] \supseteq T'$ in proposition Eqn. (4.42) can be turned into an equality yielding the equivalent proposition in Eqn. (4.43). The backward direction (4.42)$\Leftarrow$(4.43) is trivial. To prove the forward direction (4.42)$\Rightarrow$(4.43) it suffices to see that $T[n] \subseteq T'$, which is obviously the case. Altogether it follows that Eqn. (4.42) is equivalent to

$$\exists n{<}\infty \,.\, T[n] = T' \tag{4.43}$$

Now, Eqn. (4.43) implies

$$\exists n{<}\infty \,.\, \varepsilon(T[n]) = \varepsilon(T') \tag{4.44}$$

By the definition of ε in Def. 4.17 we can rewrite Eqn. (4.44) as follows:

$$\exists n{<}\infty \,.\, (\exists i{<}n.S_i{\in}C \cap P_\varepsilon(M)) = (\exists i{<}\infty.S_i{\in}C \cap P_\varepsilon(M)) \tag{4.45}$$

It is trivial that two equal events have the same probability. Therefore, it is obvious that Eqn. (4.45) implies

$$\exists n{<}\infty \,.\, \mathsf{P}_M(\exists i{<}n.S_i{\in}C \cap P_\varepsilon(M)) = \mathsf{P}_M(\exists i{<}\infty.S_i{\in}C \cap P_\varepsilon(M)) \tag{4.46}$$

Up to now, we have proven (4.41)$\Leftrightarrow$(4.42)$\Leftrightarrow$(4.43)$\Rightarrow$(4.44)$\Leftrightarrow$(4.45)$\Rightarrow$(4.46). We proceed with proving all equivalences of the above chain by completing the ring proof (4.43)$\Rightarrow$(4.44)$\Rightarrow$(4.45)$\Rightarrow$(4.46)$\Rightarrow$(4.43). We show (4.46)$\Rightarrow$(4.43) by contraposition, i.e., we prove $\neg$(4.43)$\Rightarrow$ $\neg$(4.46). We show this by contradiction, i.e., we assume $\neg$(4.43)$\wedge$(4.46) and show that this leads to a contradiction. We start with assuming that (4.46) holds. This means that we can assume an n that makes proposition (4.46) true.

We take this n as arbitrary but fixed in the sequel.

Now, we assume that (4.43) does not hold. Therefore, we can assume that the set $T[n] \neq T'$. We know that in any case $T[n] \subseteq T'$ so that we have $T[n] \subset T'$ now. We know that the difference set of T' and $T[n]$ has the following form:

$$T' \backslash T[n] = \{t \mid \exists\, n{<}i{<}\infty \,.\, t_i{\in}C,\ \forall i{\leqslant}n.t_i \notin C,\ t \in P(M)\} \tag{4.47}$$

Now, we can assume that there exists a program run $t' \in T' \backslash T[n]$. Due to basic Markov chain properties, i.e., Theorem 2.18, and due to $t'_0 = M$ we know

$$\mathsf{P}_M(S_0 = t'_0 \cap \ldots \cap S_n = t'_n) = \prod_{i=1}^{n-1} (t'_i \to t'_{i+1}) \tag{4.48}$$

Now, we know that $t'_i \to t'_{i+1} > 0$ for all $i \in \mathbb{N}_0$, because t is a program run, i.e., $t \in P(M)$. Therefore we know

$$\mathsf{P}_M(S_0 = t_0 \cap \ldots \cap S_n = t_n) > 0 \tag{4.49}$$

Now, due to $t \in T' \backslash T[n]$ and Eqn. (4.47) we can see that

$$(S_0 = t'_0 \cap \ldots \cap S_n = t'_n) \subseteq (\exists\, n < i < \infty \,.\, S_i \in C,\ \forall i \leqslant n . S_i \notin C) \tag{4.50}$$

Due to Eqns. (4.50) and (4.49) we know by the monotonicity of the probability function, compare with Lemma 2.5, that the following holds:

$$\mathsf{P}_M(\exists\, n < i < \infty \,.\, S_i \in C \cap\ \forall i \leqslant n . S_i \notin C) > 0 \tag{4.51}$$

Due to Lemma 4.10 we can restrict any event to its program executions. Therefore, Eqn. (4.51) is equivalent to

$$\mathsf{P}_M(\exists\, n < i < \infty \,.\, S_i \in C \cap\ \forall i \leqslant n . S_i \notin C \cap P_\varepsilon(M)) > 0 \tag{4.52}$$

Next we can see that

$$\begin{aligned} &\mathsf{P}_M(\exists i < \infty . S_i \in C \cap P_\varepsilon(M)) \\ &= \mathsf{P}_M(\exists i < n . S_i \in C \cap P_\varepsilon(M)) \\ &\quad + \mathsf{P}_M(\exists\, n < i < \infty \,.\, S_i \in C\ \cap\ \forall i \leqslant n . S_i \notin C \cap P_\varepsilon(M)) \end{aligned} \tag{4.53}$$

However, we have assumed Eqn. (4.46). Therefore, and due to Eqn. (4.53), we have that

$$\mathsf{P}_M(\exists\, n < i < \infty \,.\, S_i \in C \cap\ \forall i \leqslant n . S_i \notin C \cap P_\varepsilon(M)) = 0 \tag{4.54}$$

Now, Eqn. (4.54) contradicts Eqn. (4.52). Overall, this proves (4.46)⇒(4.43) correct and therefore (4.43)⇔(4.44)⇔(4.45)⇔(4.46).

Now, let us proceed with Eqn. (4.46). Due to Lemma 4.10 we can enrich any event with spurious outcomes without changing its probability so that Eqn. (4.46) can be turned into the following equivalent proposition:

$$\exists n < \infty \,.\, \mathsf{P}_M(\exists i < n \,.\, S_n \in C) = \mathsf{P}_M(\exists i < \infty \,.\, S_i \in C) \tag{4.55}$$

By the definition of hitting probabilities and bounded hitting probabilities in Def. 2.23 and Def. 2.24 we know that Eqn. (4.55) is equivalent to

$$\exists n < \infty \,.\, \eta^n_S \langle M, C \rangle = \eta_S \langle M, C \rangle \tag{4.56}$$

Now, we have completed the proof as Eqn. (4.56) adheres to the definition of $\beta(M)$, i.e., the definition of bounded termination according to Def. 4.3. $\square$

We say that a program run p vanishes if taken as an event $\varepsilon(p)$, it has a zero percent probability $\mathsf{P}_{p_0}(\varepsilon(p)) = 0$. From the terminology that we have introduced, the fact that $T(M) = P(M)$ describes the fact that all program runs of a program M terminate – see Table 4.1. If all program runs of a program M terminate we know, by definition, that the degree of termination of M equals one, i.e., $\mathsf{P}_M(T_\varepsilon(M)) = \mathsf{P}_M(P_\varepsilon(M)) = 1$. This explains the empty field in Table 4.1.

4.5 Central Graph Cover Lemmas

Given a digraph G and one of its nodes v, the node cover of v in G, or just cover of v, denoted by $Cover_G(v)$ is the set of nodes that we can reach by a walk starting in v. Equally, the path cover of v, denoted by $\pi\text{-}Cover_G(v)$ is the set of nodes that we can reach via a path starting in v.

Definition 4.23 (Graph Covers) *Given a digraph $G = \langle V_G, E_G \rangle$ and a node $v \in V_G$, we define the node cover $Cover_G(v)$ and the path cover $\pi\text{-}Cover_G(v)$ as follows:*

$$Cover_G(v) =_{DEF} \kappa(\star W_G(v)) \tag{4.57}$$

$$\pi\text{-}Cover_G(v) =_{DEF} \kappa(\pi(\star W_G(v))) \tag{4.58}$$

With respect to the reachability of nodes, infinite walks make no difference. When a node is on an infinite walk, it is also on a finite walk. Therefore, the nodes of infinite walks of a node v are already included in the nodes that we can reach by finite walks starting in v. The same is true for paths.

Lemma 4.24 (Covers of Finite Walks and Paths) *Given a digraph $G = \langle V_G, E_G \rangle$ and a node $v \in V_G$, we have that:*

$$\kappa(\omega W_G(v)) \subseteq Cover_G(v) \tag{4.59}$$

$$\kappa(\pi(\omega W_G(v))) \subseteq \pi\text{-}Cover_G(v) \tag{4.60}$$

Proof. We show Eqn. (4.59) only. Due to Def. 4.23 we need to show that

$$\kappa(\omega W_G(v)) \subseteq \kappa(\star W_G(v)) \tag{4.61}$$

For each $(w_i)_{i \in \omega} \in \omega W_G(v)$ and $n \in \mathbb{N}_0$ we have that $(w_i)_{i \in \{0,\ldots,n\}} \in n W_G(v)$. Now, Eqn. (4.61) follows from the definitions of nodes in a graph; compare with Def 2.36. $\square$

In a digraph the set of nodes that are reachable from a start node by some arbitrary walk equals the set of nodes that are reachable by some arbitrary path from that node. In order to see this, it suffices to show that two nodes that are connected by a walk are also connected by a path. Informally, this is rather easy to see. We just need to erase all cycles from a walk in order to get the desired connecting path. In order to prove it formally, we can sharpen the proposition slightly so that we can prove it by natural induction over lengths of walks. We can show that two nodes connected by a walk are also connected by a path that is at most as long as that walk. Now in the induction step we consider a walk $w_0 w_1 \dots w_n$. If the walk is already a path we are done. If it is not a path it must be possible to identify a cycle $w_i \dots w_j$ with $i, j \leqslant n$ in it. We can erase the tail $w_{i+1} \dots w_j$ of that circle from $w_0 w_1 \dots w_n$ yielding a shorter walk $w_0 w_1 \dots w_i w_{j+1} \dots w_n$. By the induction hypothesis we then know that there must exist a path that connects w_0 and w_n. We conduct the proof more accurately in Lemma 4.25 in terms of whole covers reachable from a start node.

Lemma 4.25 (Central Cover Lemma) *Given a digraph* $G = \langle V_G, E_G \rangle$ *and a node* $v \in V_G$ *we have that*

$$Cover_G(v) = \pi\text{-}Cover_G(v) \tag{4.62}$$

Proof. Each path $p \in \pi(\oplus W_G(v))$ is a walk, i.e., $p \in \oplus W_G(v)$. Therefore, $Cover_G(v) \supseteq \pi\text{-}Cover_G(v)$ follows immediately. It remains to be shown that also $Cover_G(v) \subseteq \pi\text{-}Cover_G(v)$. We do so by showing that all finite slices of the cover are subsets of the path cover. A finite slice of a cover is simply a subset of the cover resulting from all reduction walks of a given finite length. For a node v and a number n we denote the cover slice by $Cover_{G,n}(v)$ as follows:

$$Cover_{G,n}(v) = \kappa(nW_G(v))$$

Given a cover slice $Cover_n(v)$ we say that n is the reduction length of the cover slice. Now, we have that

$$Cover_G(v) = \bigcup_{n \in \mathbb{N}_0} Cover_{G,n}(v)$$

Therefore, it suffices to show that $Cover_{G,n}(v) \subseteq \pi\text{-}Cover_G(v)$ for all n. We do so by natural induction on the reduction length of cover slices.

***In Case of* n=0:** We know that $Cover_{G,0}(v)$ consists of the single node v. We know that the walk $(v_i)_{i \in \{0\}}$ is a path and therefore also $v \in \pi\text{-}Cover_G(v)$. From this $Cover_{G,0}(v) \subseteq \pi\text{-}Cover_G(v)$ follows immediately.

***In Case of* n ⩾ 1:** We can assume that $Cover_{G,n}(v) \subseteq \pi\text{-}Cover_G(v)$. In order to prove that $Cover_{G,n+1}(v) \subseteq \pi\text{-}Cover_G(v)$ we need to show that for all w and w' with $w \in Cover_{G,n}(v)$ and $\langle w, w' \rangle \in E_G$ we have that $w' \in \pi\text{-}Cover_G(v)$. For this, it suffices to show that there exists a path p' from v to w', i.e.,

$$\exists n' \in \mathbb{N}_0 \,.\, \exists p' \in \pi(n'W_G(v)) \,.\, p'_{n'} = w' \tag{4.63}$$

First, because of $Cover_{G,n}(v) \subseteq \pi\text{-}Cover_G(v)$ we know that there exists a path p from v to w, i.e.,

$$\exists n'' \in \mathbb{N}_0 \,.\, \exists p \in \pi(n''W_G(v)) \,.\, p_{n''} = w \tag{4.64}$$

We can distinguish exactly two cases now, i.e., the case that w' is different from all nodes that have appeared in the path p so far and the case that w' already appeared in p. Let's turn to the first case. From the fact that $\langle w, w'\rangle \in E_G$ and w' is different from all other nodes in p we know that the walk $p' = (p'_i)_{i\in\{0,..,n''+1\}}$ with $p'_i = p_i$ for all $i \leqslant n''$ and $p'_{n''+1} = w'$ is a path with reduction length $n''+1$ and therefore Eqn. (4.63) is satisfiable with $n' = n''+1$ in this case. Next, let us turn to the case that w' already appeared in the path p. We can assume that it appeared at reduction length $n''' \leqslant n''$. Now, we know that $p' = (p'_i)_{i\in\{0,..,n'''\}}$ with $p'_i = p_i$ for all $i \leqslant n'''$ is a path with $p_{n'''} = w'$ and reduction length n'''. Again, Eqn. (4.63) is satisfiable with $n' = n'''$. □

Theorem 4.26 (Computability of the Finite Cover) *Given a k-ary digraph $G = \langle V_G, E_G\rangle$, such that the function $v \in V_G \mapsto \{\langle v, v'\rangle \in E\}$ is total computable. Then, $Cover_G(v)$ is computable for all $v \in V$ for which $Cover_G(v)$ is finite, i.e., the following partial function is computable:*

$$v \in V \mapsto \begin{cases} Cover_G(v) & , Cover_G(v) \text{ is finite} \\ \bot & , else \end{cases} \tag{4.65}$$

Proof. First, we define a recursive algorithm *cover* that is intended to compute existing finite covers of nodes as follows:

$$cover : V_G \nrightarrow \mathbb{F}(V_G) \tag{4.66}$$

$$cover' : \mathbb{F}(V_G) \times V_G \nrightarrow \mathbb{F}(V_G) \tag{4.67}$$

$$cover(v) = cover'(\emptyset, v) \tag{4.68}$$

$$cover'(M, v) = \begin{cases} M & , v \in M \\ M \cup \{v\} & , \{\langle v,v'\rangle \in E_G\} = \emptyset \\ \bigcup\limits_{\langle v,v'\rangle \in E} cover'(M \cup \{v\}, v') & , else \end{cases} \tag{4.69}$$

Note, that, given a set S, the set $\mathbb{F}(S) \subseteq \mathbb{P}(S)$ denotes the set of finite subsets of S. We will show that $cover(v)$ terminates and correctly computes the cover of a given node v if the cover of v is finite and fails to terminate if the cover of v is infinite.

Intuitively, the intention of the algorithm is clear. The algorithm dovetails, via the helper function *cover'*, all possible walks starting from a given node

n and stops each walk whenever it encounters a node that has already been visited by this walk. This way, the algorithm steps through all possible paths starting from a node and collects all the nodes to be yielded as the final result – it is the first parameter of *cover'* that serves as an accumulator for the final result. Furthermore, if there is an infinite path, which is true in case the cover is infinite, the algorithm fails to terminate. Up to now we have described that the algorithm correctly computes finite path covers. Now, Lemma 4.25 applies. Lemma 4.25 states that the cover of a node equals its path cover. Therefore, the algorithm correctly computes finite covers.

We proceed showing both parts of the claim, i.e., first termination and correctness in case of finite covers and then, second, non-termination in case of infinite covers.

(Termination and Correctness) First, we define the following function:

$$c'(M,v) = M \cup \kappa(\{\, p \mid p \in \pi(\star W_G(v)), \kappa(p) \cap M = \emptyset\}) \tag{4.70}$$

Now, as a first step, we show that $cover'(M,v)$ terminates in case $Cover_G(v)$ is finite and then computes $c'(M,v)$ for all $M \subseteq V_G$ and nodes $v \in V_G$. Let us assume that the size of the cover, $|Cover_G(v)|$, equals n. We know that the sequence length of a path p equals its size, i.e., $\#(p) = |p|$. Therefore we know that the sequence length of each path in $\pi(\oplus W_G(v))$ must be smaller than or at most equal to n. We can exploit this to prove the supposed property of *cover'*. We conduct the proof for all nodes v that have a finite cover, by natural induction on the size n of such v.

***In Case of* n=1:** We need to consider three cases corresponding to the three clauses in the definition of *cover'* in (4.69). In case $v \in M$ we know that $cover'(M,v)$ terminates with $cover'(M,v) = M = M \cup \{v\}$. Furthermore, we then know that $\kappa(p) \cap M \neq \emptyset$ for all $p \in \pi(\star W_G(v))$ and therefore we also know that $\kappa(\{\, p \mid p \in \pi(\star W_G(v)), \kappa(p) \cap M = \emptyset\})$ is empty. From this follows that $c'(M,v) = M = M \cup \{v\}$ and therefore $cover'(M,v) = c'(M,v)$.

In the next case we have $v \notin M$ and $\{\langle v,v'\rangle \in E\} = \emptyset$. In that case we know that $cover'(M,v)$ terminates with $cover' = M \cup \{v\}$. Furthermore, we know that in this case $(v_i)_{i\in\{0\}}$ is the only path in $\star W_G(v)$. Therefore we know that $\kappa(\{p|p \in \pi(\star W_G(v)), \kappa(p) \cap M = \emptyset\})$ equals $\{v\}$ and therefore $c'(M,v)$ equals $M \cup \{v\}$ so that again $cover'(M,v) = c'(M,v)$.

Now, let us assume that the third case applies, i.e., $v \notin M$ but there exists $\langle v,v'\rangle \in E_G$. Now, due to the current induction case we know that $|Cover_G(v)| = n = 1$. Therefore, we know that the only edge $\langle v,v'\rangle \in E_G$ is the edge $\langle v,v\rangle$. Therefore, we know that $cover'(M,v) = cover'(M \cup \{v\},v)$. Now the first clause of Eqn. (4.69) applies to $cover'(M \cup \{v\},v)$, so that $cover'(M \cup \{v\},v) = M \cup \{v\}$ and therefore also $cover'(M,v) = M \cup \{v\}$. Furthermore, we know that $(v_i)_{i\in\{0,1\}} = vv$ is the only path in $\star W_G(v)$ in this case. Therefore we know that $\kappa(\{p|p \in \pi(\star W_G(v)), \kappa(p) \cap M = \emptyset\})$ equals $\{v\}$ and therefore $c'(M,v) = M \cup \{v\}$ so that again $cover'(M,v) = c'(M,v)$.

***In Case of* n⩾2:** Again, we need to consider three cases. However, the proof for the first case is completely equal to the induction case of $n = 1$ above. The second case can actually not appear in case $|Cover_G(v)| = n \geqslant 2$, because there must exist a node $v' \neq v$ that is reachable from v via an edge. Let us turn to the third case, in which we have that $v \notin M$ but $\{\langle v, v'\rangle \in E\} \neq \emptyset$. By the premises of the lemma we know that $\langle v, v'\rangle \in E$ is computable for each v'. Therefore, $cover'(M, v)$ can be computed and equals:

$$\bigcup_{\langle v,v'\rangle \in E} cover'(M \cup \{v\}, v') \tag{4.71}$$

Now, we know for all v' with $\langle v, v'\rangle \in E$ that the maximum length of its paths is reduced by one, i.e., $\#(\pi(\star W_G(v'))) < \#(\pi(\star W_G(v)))$. Therefore, by the induction hypothesis we know for each v' that $cover'(M \cup \{v\}, v')$ terminates with the value of $c'(M \cup \{v\}, v')$ so that Eqn. (4.71) equals

$$\bigcup_{\langle v,v'\rangle \in E} c'(M \cup \{v\}, v') \tag{4.72}$$

Now, due to Eqn. (4.70) we have that Eqn. (4.72) equals

$$\bigcup_{\langle v,v'\rangle \in E} M \cup \{v\} \cup \kappa(\{\, p \mid p \in \pi(\star W_G(v')), \kappa(p) \cap (M \cup \{v\}) = \emptyset\}) \tag{4.73}$$

Now, Eqn. (4.73) can be rewritten immediately as follows:

$$M \cup \bigcup_{\langle v,v'\rangle \in E} \kappa(\{\, v \bullet p \mid p \in \pi(\star W_G(v')), \kappa(p) \cap (M \cup \{v\}) = \emptyset\}) \tag{4.74}$$

Now, we know that $v \bullet p$ is a path if and only if p is a path and v does not occur in p, i.e., $v \notin \kappa(p)$. Therefore, we can rewrite Eqn. (4.74) as follows:

$$M \cup \kappa(\{\, v \bullet p \mid v \bullet p \in \pi(\star W_G(v)), \kappa(p) \cap M = \emptyset\}) \tag{4.75}$$

Now, we can again exploit the fact that $v \notin M$, which holds in the currently considered case. Therefore, we have that Eqn. (4.75) equals

$$M \cup \kappa(\{\, q \mid q \in \pi(\star W_G(v)), \kappa(q) \cap M = \emptyset\}) \tag{4.76}$$

Finally, due to the definition of c', see Eqn. (4.70), we know that Eqn. (4.76) equals $c'(M, v)$.

Now, we have finished the induction proof and know that $cover'(M, v)$ terminates and equals $c'(M, v)$ for all v that have a finite cover. Next, by the definition of c', we know that the following holds for all nodes:

$$c'(\emptyset, v) = \emptyset \cup \kappa(\{\, p \mid p \in \pi(\star W_G(v)), \kappa(p) \cap \emptyset = \emptyset\}) = \pi(\star W_G(v)) \tag{4.77}$$

Given Eqn. (4.77) we have already shown that *cover* correctly computes the path cover $\pi(\star W_G(v))$ for v that have a finite cover. Now, by the central cover Lemma 4.25 we know that the cover of a node equals its path cover. Therefore we also know that *cover* correctly computes the cover $\star W_G(v)$ for all v that have a finite cover.

(Non-Termination) To complete the proof, we need to show that the algorithm *cover* fails to terminate in case of infinite covers. Given a node v so that $Cover_G(v)$ is infinite. Then, due to the fact that G is k-ary, by König's Lemma, see Lemma 2.39, we know that there exists an infinite path $\omega \in \pi(\omega W_G(v))$ starting from $v = \omega_0$. For ω we know that ω_{i+1} is not an element of the set $\{\omega_0, \ldots, \omega_i\}$ for all i, because ω is a path. Therefore, there exists the following infinite series of recursive calls, which amounts to the non-termination of $cover(v)$:

$$cover'(\emptyset, \omega_0) \rightsquigarrow cover'(\{\omega_0\}, \omega_1) \rightsquigarrow \ldots \rightsquigarrow cover'(\{\omega_0, \ldots, \omega_{i-1}\}, \omega_i) \rightsquigarrow \ldots \tag{4.78}$$

□

Let us proceed with some remarks on the proof conducted for Theorem 4.26 and the definitions it establishes. First, the notation $f \rightsquigarrow f'$ in Eqn. (4.78) is an ad hoc pseudo-notation that indicates that a function call f leads to a further function call f'. Then, the condition $\{\langle v, v' \rangle \in E\} = \emptyset$ in the second clause of Eqn. (4.69) can be rewritten in more convenient form as $\nexists v'.\langle v, v' \rangle \in E$. It has got the chosen form just because it equals the big union comprehension in the third clause of the definition this way. For the application of our lemma to the reduction graph later, the second clause is even non-relevant, because in our reduction graph, each node has at least one outgoing node. For the general case of arbitrary digraphs, the second clause is indispensable. The function c' yields the nodes of all paths that do not contain a node from M. Note that the definition of $cover'$ does not exclude nodes from a path just because it extends to a longer path that eventually contains some nodes from M. This means that if we have a path $a_1 \cdots a_i a_m a_{i+2} \cdots$ with $a_m \in M$ but $a_1, \ldots, a_i \notin M$ we still have that $a_1, \ldots, a_i$ belong to $c'(a_1, v)$ just because $a_1 \cdots a_i$ is also a path in $\star W_G(v)$.

4.6 Path Stoppability

In this section we introduce a broadened notion of termination for programs of the probabilistic lambda calculus, which we call path stoppability. It relies on the notion of term covers and characterizes a class of programs for which we can determine the degree of termination, albeit they might possess some non-terminating program runs. A term cover of a program is simply the set of terms that can be reached by any of its program runs. We use the concepts introduced in the preceding sections to formally define the concept of term covers. Term covers are just graph covers of the reduction graph R. Given a program M, we define its term cover as its graph cover $Cover_R(M)$ with M taken as start node. Often, we talk of the term cover of a program simply as its cover. Occasionally, we also drop the name of the reduction relation and denote a term cover $Cover_R(M)$ simply as $Cover(M)$. We do so, e.g., in Table 4.2.

Let us turn to the central result. The degree of termination is computable for programs with finite term covers. The algorithm consists of three steps:

- Compute the finite term cover, if it exists.
- Generate a linear equation system that describes the hitting probabilities of ever hitting a constant for all involved terms from the finite cover. A finite linear equation system can always be found in case of finite covers.
- Solve the linear equation system to find its least solution. In case of a finite equation system, its least solution can always be determined algorithmically by Gauss-Jordan elimination.

The computation of finite covers in the above result is just an application of the computability of graph covers, see Theorem 4.26, to the case of the reduction graph. Then, the computability of termination degrees is just a consequence of Gauss-Jordan elimination. Next, it is the task of Theorem 4.27 to give a formal account of the computability of termination degrees.

Theorem 4.27 (Computability of Termination Degrees) *The degree of termination is computable for all programs that have a finite term cover, i.e., the following partial function is computable:*

$$\delta = M \in \Lambda_P \mapsto \begin{cases} \eta_S\langle M, C\rangle & , Cover_R(M) \textit{ is finite} \\ \perp & , else \end{cases} \tag{4.79}$$

Proof. Once more, note that $\eta_S\langle M, C\rangle$ equals $\mathsf{P}_M(\varepsilon(T(M)))$. By Theorem 4.26 we know that $Cover_R(M)$ is computable in case it is finite. Therefore, as the first step we compute $Cover_R(M)$. Next, by Def. 4.1 and Theorem 2.29 we know that the vector of termination degrees $(\eta_S\langle M, C\rangle)_{M\in\Lambda_P}$ is the least solution of the following infinite equation system:

$$\begin{aligned} \forall c \in C \,.\, \eta_S\langle c, C\rangle &= 1 \\ \forall N \notin C \,.\, \eta_S\langle N, C\rangle &= \sum_{N'\in\Lambda_P} (N \to N') \cdot \eta_S\langle N', C\rangle \end{aligned} \tag{4.80}$$

Due to the fact that $Cover_R(M)$ is finite we can reduce the equation system (4.80) to an equivalent finite version. To see this, we bring the equation system (4.80) into a different form, now consisting of four subsystems:

$$\forall c \in C \backslash Cover_R(M) . \eta_S\langle c, C\rangle = 1 \tag{4.81}$$

$$\forall N' \in \Lambda_P \backslash C \backslash Cover_R(M).\eta_S\langle N', C\rangle = \sum_{N'' \in \Lambda_P} (N' \to N'') \cdot \eta_S\langle N', C\rangle \quad (4.82)$$

$$\forall c \in Cover_R(M) \cap C.\eta_S\langle c, C\rangle = 1 \quad (4.83)$$

$$\forall N \in Cover_R(M) \backslash C.\eta_S\langle N, C\rangle = \sum_{\substack{N' \in \\ Cover_R(M)}} (N \to N') \cdot \eta_S\langle N', C\rangle + \sum_{\substack{N' \notin \\ Cover_R(M)}} (N \to N') \cdot \eta_S\langle N', C\rangle \quad (4.84)$$

The subsystems (4.81) and (4.82) single out equations that give hitting probabilities for programs that are not in the cover. Subsystems (4.83) and (4.84) contain the remaining equations. Now, consider subsystem (4.84). Here the second sum, which iterates over programs that are not in the cover, sums up to zero, because each single summand is zero. To see this, consider the probability $N \to N'$ in the sum. We know that for each of the summands, N is in the cover, whereas N' is not. Now, the fact that N is in the cover implies that there is a path $Mt_1t_2 \cdots t_{n-1}N$ from M to N. Now, if $N \to N'$ were different from zero there would exist a path $Mt_1t_2 \cdots t_{n-1}NN'$ from M to N', which we know is not true due to the iterator condition $N' \notin Cover_R(M)$ of the sum. Therefore, $N \to N'$ must be zero. Therefore, we can drop the second sum from all equations in subsystem (4.84). Furthermore, we see that all left hand side variables of subsystems (4.81) and (4.82) are neither directly relevant to the eventual object of interest, which actually consists only of a single variable, i.e., $\eta_S\langle M, C\rangle$, nor do they occur in any of the equations of the subsystems (4.83) and (4.84). Therefore, also subsystems (4.81) and (4.82) can be dropped without loss for the determination of the final solution. Therefore, eventually we can transform the equation system (4.80) into the following equivalent equation system:

$$\begin{aligned} &\forall c \in Cover_R(M) \cap C.\eta_S\langle c, C\rangle = 1 \\ &\forall N \in Cover_R(M) \backslash C.\eta_S\langle N, C\rangle = \sum_{\substack{N' \in \\ Cover_R(M)}} (N \to N') \cdot \eta_S\langle N', C\rangle \end{aligned} \quad (4.85)$$

Now, due to the fact that $Cover_R(M)$ is finite we also know that the equation system (4.85) is finite. Now, for each finite linear equation system, it is possible to compute the least solution vector, e.g., by transformation to row echelon form by application of the Gauss-Jordan algorithm; see, e.g., [43, 189]. □

The class of programs that have a finite cover defines a widened notion of termination that can also be characterized in terms of the core algorithm used to determine the cover of programs; see Def. 4.28. We coin the term path stoppability, also called p-stoppability, for this concept. Where the existence of a finite cover is a rather declarative notion, the notion of p-stoppability is a rather operational notion. Obviously, the notion of p-stoppability and the existence of a finite cover are equivalent, as expressed by Lemma 4.29. A program is path stoppable, or p-stoppable, in case it has a finite cover.

Definition 4.28 (Path Stoppability) *A program M is p-stoppable* **iff** *applying the path-stopping algorithm π to M terminates, i.e., if $\pi(M)$ terminates, where π is defined as follows:*

$$\pi \; : \; \Lambda_P \twoheadrightarrow \mathbf{1} \tag{4.86}$$

$$\pi' \; : \; \mathbb{F}(\Lambda_P) \times V_G \twoheadrightarrow \mathbf{1} \tag{4.87}$$

$$\pi(A) = \pi'(\emptyset, A) \tag{4.88}$$

$$\pi'(A, v) = \begin{cases} \underline{1} & , M \in A \\ \bigcup\limits_{\langle M, M' \rangle \in E} \pi'(A \cup \{M\}, M') & , v \notin A \end{cases} \tag{4.89}$$

Corollary 4.29 (Path Stoppability and Finite Covers) *A program M is path stoppable, i.e., $\pi(M)$ terminates,* **iff** *its cover $Cover_R(M)$ is finite.*

Proof. Corollary from Def. 4.28 and Theorem 4.26. Basically, the algorithm $\pi \; : \; \Lambda_P \twoheadrightarrow \mathbf{1}$ in Def. 4.28 mimics the algorithm $cover \; : \; V_G \twoheadrightarrow \mathbb{F}(V_G)$ in Theorem 4.26. □

Now that we have coined the term path stoppability for programs, we also use it for single program runs. We call those program runs path stoppable that are not paths, i.e., that contain a cycle; see Def. 4.30. The intuition behind this should be clear. Each walk that is not a path is eventually stopped by the algorithm *cover* in Theorem 4.26. It is executed by *cover* as long as it is a path, i.e., its longest path prefix is executed. As soon as *cover* discovers that one of the program executions that it maintains is a cycle, it stops this program execution. This also explains why we have chosen to name the established notion of stoppability as path stoppability. If all program runs eventually run into cycles during simultaneous execution by *cover*, the whole program is stoppable. This is also expressed by the following easy Lemma 4.31.

Definition 4.30 (Path Stoppability of Program Runs) *A program run $p \in P(M)$ is path stoppable, also called p-stoppable,* **iff** *p is not a path in the reduction graph, i.e., $p \notin \pi(P(M))$.*

Corollary 4.31 (Path Stoppability of Programs) *A program M is path stoppable* **iff** *all of its program runs are path stoppable.*

Proof. Corollary from Defs. 4.28 and 4.30 and Theorem 4.26. □

We proceed with the fact that a program with an infinite cover necessarily possesses a non-terminating program run, as expressed in Lemma 4.32. Remember that Lemma 4.32 is needed as one of the essential Lemmas in our analysis of termination behavior; compare with Table 4.3 and Sect. 4.1.2.

Lemma 4.32 (Infiniteness of Covers and Non-Termination) *If a program M has an infinite cover then it has a non-terminating program run.*

Proof. If a program M has an infinite cover we know by König's Lemma, i.e., Lemma 2.39, that M has an infinite path, i.e., there exists $q \in \pi(P(M))$. Due to Corollary 4.21 we know that all $p \in \pi(P(M))$ are non-terminating and therefore also p is non-terminating, i.e., $p \notin T(M)$. □

Path Computability

We can adapt the result from Theorem 4.27 to the computation of reduction probabilities $M \Rightarrow N$. For this purpose, we define the partial function $\Rightarrow_p$ for arbitrary programs M and N, called path computation of $M \Rightarrow N$ or p-computation for short, as follows in Def. 4.33.

Definition 4.33 (Path Computation) *We define the partial function* $\_ \Rightarrow_p \_ : \Lambda_P \times \Lambda_P \to [0,1]$*, called path computation or p-computation, for all program M and N as follows:*

$$M \Rightarrow_p N \mapsto \begin{cases} M \Rightarrow N & , Cover_R(M) \textit{ is finite} \\ \bot & , else \end{cases} \tag{4.90}$$

Along the lines of Theorem 4.27 we have that also path computation is a computable function, which justifies its name. As in the argumentation preceding Theorem 4.27 the computation is about determining the finite term cover of M, if it exists, and then generating and solving an appropriate linear equation system.

Path Stoppability and Program Analysis

Path stoppability is a form a program analysis. It can also be, somehow, related to the field of static program analysis [262]; compare also with [71, 198, 199, 53] for some important examples from the field of probabilistic computation. In the field of static program analysis we try to figure out useful properties of a program before, or let's say independent of, run-time. Properties might be measures from the field of software quality, i.e., code quality. In principle, all kinds of non-functional requirements, i.e., concerning performance, interoperability, maintainability or, in particular, IT security, might be subject to static program analysis. However, also functional requirements, i.e., correctness criteria, might be checked with static program analysis. Usually, we would rather expect that the properties investigated by static program analysis are decidable properties. Then, a good approach is to characterize the field of static program analysis in terms of compiler construction [5, 216] terminology. Then, static program analysis might be defined as the field of context-sensitive as opposed to context-free program properties. Then, the type system of a language would be an important, albeit reductionist example of static program analysis. However, all that said, we are not sure whether decidability should be considered a necessary ingredient of static program analysis at all. In any case, note that path stoppability is a semi-decidable notion. In that sense, we would not say that path stoppability is an instance of static program analysis. Nevertheless, it is a form of program analysis, and as such, it is a very important one, both theoretically and practically.

Path stoppability is a severe notion of program analysis. It takes termination behavior as its subject of investigation. It defines a class of probabilistic programs that might be trapped in a non-terminating program run and provides an algorithm for their identification and for the calculation of their termination degrees. The fact that path stoppability is semi-decidable does not hurt too much, just because the subject of investigation is termination, or to say it even better, non-termination. This can be seen best in the special case of deterministic programs. Of course, path stoppability can also be applied to deterministic programs. Let us assume that you apply the *cover* algorithm to a non-terminating program, let us call it the object program, before you actually run the program. If the object program run has a finite cover, then the *cover* algorithm will stop and you have won something, because you will not try to execute the object program with its original operational semantics any more. If the object program has an infinite cover, of course, the *cover* algorithm will not terminate any more due to its semi-decidability. But this does not hurt, because you have not lost anything: the object program with its original operational semantics would also not terminate.

Now, a similar argument as in the special case of deterministic programs can be constructed for the general case of probabilistic programs. Here, conducting the test for path stoppability in advance, before we actually execute the object program, can have the effect that we lose some outcomes in case of a program with infinite cover. We will wait for the termination of the *cover* algorithm forever; however, if we had executed the object program we might have observed a positive outcome. Actually, maybe the object program even has a termination degree of one. Remember that we have seen an example of a program with termination degree one that has a non-terminating program run. In such a case the chance that we miss an outcome is a 100 per cent.

However, executing the program analysis in advance is just a wrong, or let's say misleading, thought experiment. We have been guided by our preoccupation with static program analysis. Still, it makes sense to conduct the path stoppability analysis independent of the actual program execution. Let's just assume that we conduct the program analysis in parallel with the actual program analysis. Let's neglect the extra computing resources for doing so. In practice, of course, we cannot neglect them, in particular, because the path stoppability analysis actually requires significant resources. However, if we neglect the necessary extra resources, the semi-decidability of our program analysis again does not hurt, as in the case of deterministic programs described above. If the analysis terminates, we can stop the whole execution. If the analysis does not stop, it does not hurt.

Now, if the task is to investigate the degree of termination of a given program, executing the program analysis does not hurt at all, even if we also take into account the needed computing resources. In any case, we need to simulate on the basis of the given program, i.e., we have to dovetail its program runs. So, in that scenario the parallel execution of the path stoppability analysis comes at no price, it can be masked into the running simulation in that case.

All of the above discussion is, of course, quite narrative. But we hope that it already shows the practical relevance of path stoppability and gives an impression of its use cases.

Strength of Path Stoppability

The class of programs characterized by path stoppability is quite large. And it can be considered as strong, because not only does it yield information about the termination behavior of a program, but its result can also be exploited to determine the degree of termination of the program, of course, only in case that the program analysis terminates. In a sense, it can be considered as strong, because the *cover* algorithm works for the largest class of programs for which Gauss-Jordan elimination can be exploited, i.e., the class of all programs that have finite term cover. Of course, the notion of termination can be further broadened, arbitrary kinds and arbitrarily many possible extensions are imaginable, and among those, there are surely many very practical notions. Here, we enter the field of abstract interpretation and related disciplines. However, the class of program characterized by path stoppability is particularly natural.

It is always the same with program analysis. To get the point take the brutal case of total termination as a program analysis property. This means we are interested in analyzing whether a program terminates. Here, it is a minor point whether we are interested in the question for all of its possible input data or some, or even a single input datum. Now, as we all know, total termination is not decidable, i.e., there is no algorithm that always terminates and correctly analyses whether the given object always terminates. However, this should not demotivate us to find better and better analyses for program termination. This means that if we cannot find the absolute algorithm it nevertheless makes sense to implement approximations to it. For example, in case of termination analysis, it adds value to parse a program for obvious endless loops. By data flow analysis we can find better and better means to detect endlessly looping call structures.

The discussion that we just conducted is an informal discussion. All of the crucial categories that we have used in the above discussion to describe program analysis such as size, strength, optimality etc. are necessarily vague.

4.7 Program Reduction Trees

The reduction trees introduced in this section are a natural notion. Reduction trees are a common device in the analysis of reduction systems. They are so common that they are often introduced directly, without mention of the notion of reduction graph. However, for us, the distinction of reduction graphs and reduction trees is important, because the two structures are used for different purposes of program analysis.

4.7.1 Intuition Behind the Reduction Tree Construction

We will define a redution tree, denoted by $\tau[M]$, for each program M. A reduction tree $\tau[M]$ turns reduction walks in R into reduction paths. If a reduction walk contains a cycle from a term N back to N, a new node is generated for each repeated occurrence of N in this walk. To see how this works have a look at Fig. 4.1. The left-hand side of the figure shows the reduction graph of $\mu\lambda x.x|0$, or, to be more precise, the part of the reduction graph R that contains all walks starting from $\mu\lambda x.x|0$. The right-hand side shows the corresponding reduction tree $\tau[M]$. In the reduction graph each term M uniquely identifies a node of the graph. This is not so in the reduction tree. Here, the terms are rather the label or let us say the content of a node. The term $\mu\lambda x.x|0$, the term $(\lambda x.x|0)|\mu\lambda x.x|0$ and the term 0 all appear infinitely many times in the tree, however, each appearance of one of these terms stands for a distinct node. It would be possible to distinguish the nodes explicitly in the tree, e.g., by using indices like $[\mu\lambda x.x|0]'$, $[\mu\lambda x.x|0]''$, $[\mu\lambda x.x|0]'''$ and so forth for the single occurrences of $[\mu\lambda x.x|0]$. However, it is usual to save such extra efforts when a path is shown. Nevertheless, when we step from graph to walks it is our task to equip the terms with the necessary extra identities.

As we have said, reduction trees are a usual means to visualize or analyze the reductions of a reduction system. If a deterministic notion of reduction is fixed for a reduction system, its reduction trees are single paths. This is so whenever we fix a concrete evaluation strategy for the lambda calculus or a functional programming language and this is also so for PCF [223]. However, in general, the reduction to a term is not unique in a reduction system. Then it makes sense to consider the set of possible reductions as a tree. Usually, the reduction tree is introduced directly, without the need to introduce or even to mention a notion of reduction graph. Then, it is taken as implicit that occurrences of terms in example paths stand for different nodes. It might not even be worth mentioning, because this somehow follows from the fact that the considered objects are paths, and not arbitrary walks. However, for us, the distinction is important. The notion of reduction tree would not have been appropriate for us to prove the propositions of Sect. 4.6 on path stoppability of programs. The notion of path stoppability relies on the distinction between reduction walks, which may have cycles, and reduction paths, which do not have cycles. However, in a reduction tree, all walks are paths. On the

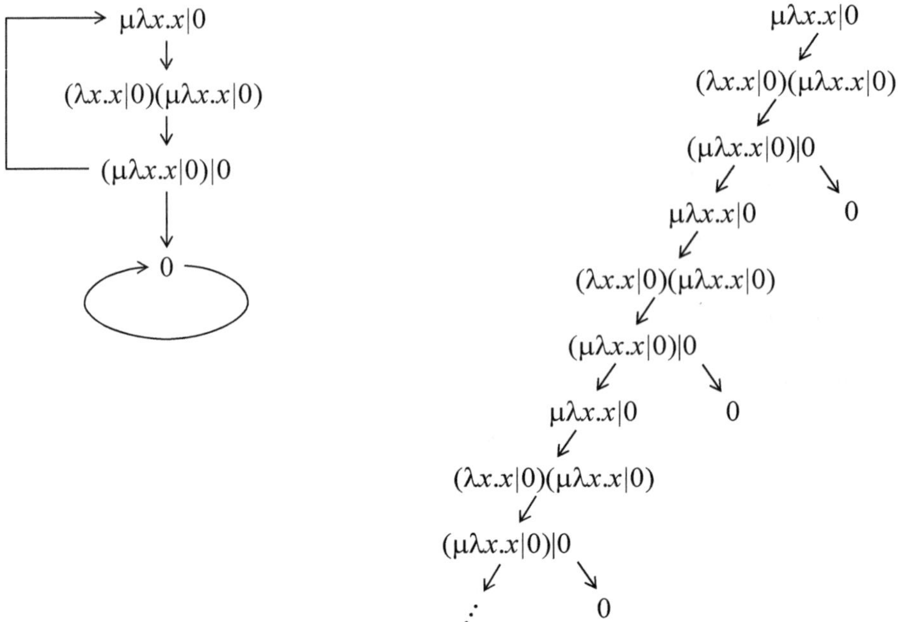

Fig. 4.1. Reduction graph and reduction tree compared

other hand, we need the notion of reduction trees to prove other properties of program behavior for which the reduction graph is not the optimal tool. For example, we will exploit the notion of reduction tree to prove that every unbounded program necessarily has a non-terminating program run in Lemma 4.39.

4.7.2 Program Runs in Reduction Graphs and Trees

The most basic difference between reduction graphs and reduction trees is that all walks in the reduction tree are paths. Of course, cycles in reduction walks can be maintained in the reduction tree, because they can be retrieved from the terms contained in nodes, however, in the reduction graph, walks that are not paths are immediately distinguished from paths. Now, there is another crucial difference between reduction graphs and reduction trees. It is about how program runs are represented in a reduction graph and how they are represented in a reduction tree. In a reduction graph, all program runs are represented by the infinite walks in the graph. Terminating program runs are those that eventually reach a constant and non-terminating program runs are those that never reach a constant. Nevertheless, both the terminating and the non-terminating program runs are modeled as infinite walks. The finite walks in the graph are never considered as program runs themselves, they always only serve as approximations to program runs. This is different in program trees. Here, all walks are paths. The infinite paths in a reduction tree are exclusively used to represent non-terminating program runs. Consequently, a terminating program run always corresponds to a finite path in the reduction tree. This one-to-one identification of termination with finiteness strongly meets our intuition of program behavior and makes the strength of reduction trees in proofs. To see this correspondence, once more have a look at the program example in Fig. 4.1. The program $\mu\lambda x.x|0$ has two kinds of program runs, one non-termination program run that is kept in a cycle always returning to $\mu\lambda x.x|0$, and infinitely many program runs that terminate with zero. The non-terminating program run is turned into the unique infinite path in $\tau[M]$ seen as the spine of the tree in Fig. 4.1. The infinitely many terminating program runs are turned into infinitely many finite paths in $\tau[M]$ each ending with a node that contains 0 as term.

We have seen earlier that the notion of program run has different representations. We have defined program runs in the σ-algebra, which is an event. A program run in the σ-algebra is a largest set of program executions that cannot be distinguished further only by means of the Markov chain semantics. Then, a program run in a graph is an infinite sequence of terms and is therefore an infinite walk in the reduction graph. Now we have a third representation of program run, i.e., a certain kind of path in the program reduction tree. To avoid confusion we usually talk about a tree program run if we mean the presentation of a program run in a tree, whereas we rather often talk about a program run in a graph simply as a program run.

4.7.3 Construction of the Reduction Tree

In order to equip terms with the necessary extra identities to become tree nodes, we simply take the finite walks of the reduction graph R as nodes in reduction trees. However, not all of the walks of the reduction walk are included in the set of nodes of the reduction tree. This is so, because we want to anticipate the strict match between terminating program runs and finiteness of corresponding paths in the tree. Therefore, only those finite walks in which a constant appears at most once are taken as nodes of a reduction tree.

Given a program M the nodes of the reduction tree $\tau[M]$ are the walks in R starting with M and containing a constant at most once. Two kinds of walks are taken as nodes. The first kind of walk does not contain a constant. The second kind of walk does contain a constant, but at most once. If a finite walk contains a constant at most once, this constant must necessarily be the last term of the walk. The walks that contain a constant at most once will serve as leaves in reduction trees, as can be seen in Fig. 4.1. The walks that contain no constant will always be inner nodes of a reduction tree.

Each node in the reduction tree is a walk in the reduction graph. Given a node v in the reduction tree, we call its least element M, i.e., the least element of the tree node taken as a graph walk, the term contained in v or the term represented by v. We then also simply say that M is the term of v. Now, we define that an edge between two tree nodes v and v' exists if and only if there exists an edge between the two terms contained in v and v' in the graph R or, equally, if and only if there exists a transition between the terms contained in v and v' in the one-step semantics.

Now, given a path in a reduction tree, we say that the path contains a term M if the term M is contained by one of its nodes. Furthermore, we call the term contained in the last node of a path the term reached by this path. Of course, the terms reached by paths are the leaves of the tree.

Definition 4.34 (Program Reduction Tree) *Given a program M we define its program reduction tree, denoted by $\tau[M]$, as the rooted directed tree $\tau[M] = \langle V, E, r\rangle$ as follows:*

$$V \subseteq \star W_R(M) \tag{4.91}$$

$$V = \{m_0 \cdots m_n \in \star W_R(M) \mid \forall i < n \,.\, m_n \notin C\} \tag{4.92}$$

$$E = \{\langle m_0 \cdots m_n \,,\, m_0 \cdots m_n m_{n+1}\rangle \mid \langle m_n, m_{n+1}\rangle \in E_R\} \tag{4.93}$$

$$r = (M)_{i\in\{0\}} \tag{4.94}$$

Actually, the definition of reduction trees in Def. 4.34 must be shown to be well defined. The definition contains the implicit proposition that the structure $\langle V, E, r\rangle$ it defines is actually a rooted tree. Obviously, the structure $\langle V, E\rangle$ is a digraph and r is a designated element in this digraph. Now, in order to show that $\tau[M]$ is a rooted tree with r as root, it remains to show that $\tau[M]$ is acyclic and each node $v \in V$ which is different from the root is connected to the root by a unique path. This can be seen by the definition of the edges E and can be proven, formally, by induction on the length of walks underlying the nodes in V.

As we have described in Sect. 4.7.2 we now define tree program runs, terminating program runs and non-terminating tree program runs. Given a program M its non-terminating tree program runs are the infinite paths in its reduction tree starting from M, whereas its terminating program runs are the finite paths in the reduction tree starting from M that contain a constant. Now, the program runs of M consist of its terminating plus its non-terminating program runs.

First, note that it suffices to describe the terminating program runs as paths that contain a constant, i.e., it is not necessary to say that they contain a constant at most once. By the definition of nodes and edges of reductions it is already guaranteed that a path that contains a constant does not contain that constant more than once. Furthermore, note that it is necessary to define the set of program runs explicitly as the union of terminating and non-terminating program runs. All of the infinite paths in the tree represent a program run, whereas finite tree paths that contain no constant can be considered as approximations to program runs.

Let us turn to the definition of the several kinds of tree program runs in Def. 4.35. Recall that we use $[s]^{\blacktriangleright}$ to denote the last element of a sequence s.

Definition 4.35 (Tree Program Runs) *Given a program reduction tree $\tau[M] = \langle V, E, r\rangle$ we define its tree program runs, denoted by $P_\tau(M)$, its terminating tree program runs, denoted by $T_\tau(M)$ and its non-terminating tree program runs, denoted by $\overline{T}_\tau(M)$ as follows:*

$$T_\tau(M) = \{v_0 \cdots v_n \in \star W_{\tau[M]}(M) \mid [v_n]^{\blacktriangleright} \in C\} \tag{4.95}$$

$$\overline{T}_\tau(M) = \omega W_{\tau[M]}(M) \tag{4.96}$$

$$P_\tau(M) = T_\tau(M) \cup \overline{T}_\tau(M) \tag{4.97}$$

Definition 4.36 (Terms of a Reduction Tree) *Given a program reduction tree $\tau[M] = \langle V, E, r\rangle$, one of its nodes $v \in V$ and one of its terminating program runs $t \in T_\tau(M)$. The term of the node v is defined as its last element $[v]^{\blacktriangleright}$. The term reached by t is defined as the term of its last node $[[t]^{\blacktriangleright}]^{\blacktriangleright}$.*

By their design, the tree program runs $T_\tau(M)$, $\overline{T}_\tau(M)$ and $P_\tau(M)$ stand in a one-to-one correspondence to their program run counterparts $T(M)$, $\overline{T}(M)$ and $P(M)$. Next, we make this correspondence explicit in two mutually inverse functions $\varsigma : P(M) \to P_\tau(M)$ and $\varsigma^{-1} : P_\tau(M) \to P(M)$ in the following definition Def. 4.37 and Corollary 4.38. Note that in Def. 4.37 we use $s \bullet t$ to denote the concatenation of sequences s and t.

Definition 4.37 (Correspondence of Reduction Graphs and Trees) *Given a program M, we define $\varsigma : P(M) \to P_\tau(M)$, the interpretation of program runs as tree program runs, and $\varsigma^{-1} : P_\tau(M) \to P(M)$, the interpretation of tree program runs as program runs as follows:*

$$\varsigma(w) = \begin{cases} (w_0 \cdots w_i)_{i \in \{\,0\,,\,\ldots\,,\,\infty\,\}} & , w \in \overline{T}(M) \\ (w_0 \cdots w_i)_{i \in \{\,0\,,\,\ldots\,,\,\sqcap\{i \mid w_i \in C\}\,\}} & , w \in T(M) \end{cases} \tag{4.98}$$

$$\varsigma^{-1}(w) = \begin{cases} ([w_i]^{\blacktriangleright})_{i \in \mathbb{N}_0} & , w \in \overline{T}_\tau(M) \\ ([w_i]^{\blacktriangleright})_{i \in \{\,0\,,\,\ldots\,,\,\#(w)-1\,\}} \bullet ([[w]^{\blacktriangleright}]^{\blacktriangleright})_{i \in \mathbb{N}_0} & , w \in T_\tau(M) \end{cases} \tag{4.99}$$

Corollary 4.38 (Correspondence of Reduction Graphs and Trees) *Given a program M, the program run interpretation $\varsigma : P(M) \to P_\tau(M)$ and tree program run interpretation $\varsigma^{-1} : P_\tau(M) \to P(M)$ are inverse functions, i.e.,*

$$\varsigma \circ \varsigma^{-1} = \mathbf{id}_{P(M)}$$
$$\varsigma^{-1} \circ \varsigma = \mathbf{id}_{P_\tau(M)}$$

Proof. Omitted.

4.8 Characteristics of Bounded Termination

In this section we finish our analysis of termination behavior. We investigate how bounded termination relates to the existence of terminating program runs. We provide the missing propositions to complete the whole picture of program analysis outlined in Sect. 4.1.1; see Table 4.3. We benefit from all the technical apparatus that we have introduced in the previous sections such as bounded termination, program runs in the reduction graph, program runs in the reduction tree, and the several mutual correspondences.

A program terminates bounded if the length of its terminating program runs is capped by a maximum length and unbounded otherwise. Now, we have the following. A program that terminates unbounded necessarily has a non-terminating program run. Or, to state the contraposition, we have the following. A program that has only terminating program runs necessarily terminates bounded. The opposite of that fact does not hold in general. However, for the special case of a termination degree of one, we actually have that the opposite holds. Therefore, in case a program has a termination degree of one the above implications turn into equivalences. Then, the program has a non-terminating program run if and only if it terminates unbounded. Or, to say it differently, the program only has terminating program runs if and only if it terminates bounded. This section is exactly about these two facts, treated in Lemma 4.39 and Lemma 4.40.

Lemma 4.39 (Unbounded Termination and Non-Termination) *If a program M terminates unbounded* **then** *M has a non-terminating program run.*

Proof. We assume that the reduction tree of M is given as $\tau[M] = \langle V, E\rangle$. We assume that the program terminates unbounded. We define an embedding $v : \mathbb{N}_0 \hookrightarrow V$ which proves V to be infinite. Due to the fact that M terminates unbounded, we know that for all $i \in \mathbb{N}_0$ there exists a terminating program run $t_i = m_0 m_1 \cdots m_{i-1} m_i \cdots c\, c\, c \cdots$ with $m_j \notin C$ for all $j < i$ in the reduction graph. For the reduction graph, this fact alone does not imply the infiniteness of the reduction graph, because it is not ensured that enough of the m_js are distinct. However, in the reduction tree $\tau[M]$, the existence of the program run $t_i = m_0 m_1 \cdots m_{i-1} m_i \cdots c\, c\, c \cdots$ amounts to a new, distinct node $v(i) = m_0 m_1 \cdots m_{i-1} m_i$ for each $i \in \mathbb{N}_0$. To see this, note that the existence of the program run $m_0 m_1 \cdots m_{i-1} m_i \cdots c\, c\, c \cdots$ implies that $m_0 m_1 \cdots m_{i-1} m_i$ is a walk in the reduction graph. Therefore, due to Def. 4.34 the graph walk $m_0 m_1 \cdots m_{i-1} m_i$ is a node in the reduction tree. Furthermore, $m_0 m_1 \cdots m_{i-1} m_i$ must be different from all $v(j)$ for all $j < i$, simply because it is longer than all of these nodes taken as sequences. Therefore, it follows that v is injective so that the reduction tree $\tau[M]$ is shown to be infinite. Now, it follows by Beth's Tree Theorem 2.50 that there exists an infinite path ω in $\tau[M]$. Due to the existence of the infinite path ω we know that there exists a non-terminating program run $\varsigma^{-1}(\omega)$. □

Of course, Lemma 4.39 can also be proven by contraposition. We give an outline. If a program M has no non-terminating program run, the reduction tree $\tau[M]$ has no non-terminating tree program run due to Corollary 4.38, which means that it has no infinite path. Therefore, due to Beth's Tree Theorem, see Theorem 2.50, $\tau[M]$ has only finitely many nodes. The finite number of nodes of $\tau[M]$, let us say n, is also an upper bound of the length of paths of $\tau[M]$, because a path can contain each node at most once. All walks in a tree are paths and so are the tree program runs. This means that n is also an upper bound for the tree program runs of $\tau[M]$. Each terminating program run $t \in T(M)$ is mapped to a tree terminating program $\varsigma(t)$. Due to Def. 4.37 we know that $t \in T(M)$ is as most as long as $\varsigma(t)$. Therefore, the number of tree nodes n is also an upper bound for the length of terminating program runs, which means that program M terminates bounded; compare with Def. 4.3.

Lemma 4.40 (Degree-One and Unbounded Termination) *Given a program M. If M has termination degree of one* **and** *M has a non-terminating program run* **then** *M terminates unbounded.*

Proof. We provide a proof sketch first. By currying we can rewrite the proposition also as

$$\eta_S\langle M,C\rangle = 1 \implies (\exists t \in \overline{T}(M) \Rightarrow \neg\beta(M)) \tag{4.100}$$

Overall, we assume $\eta_S\langle M,C\rangle = 1$. Then, we will prove $\exists t \in \overline{T}(M) \Rightarrow \neg\beta(M)$ by contradiction, i.e., we will prove $\exists t \in \overline{T}(M) \wedge \beta(M)$ to be impossible.

We assume that the program has a termination degree of one. Furthermore, we assume that the program terminates bounded. Therefore, we know that there exists a bound $n \in \mathbb{N}_0$ such that $\eta_S^n\langle M,C\rangle = \eta_S\langle M,C\rangle$ and therefore $\eta_S^n\langle M,C\rangle = 1$. Be Def. 2.23 we have that $\eta_S^n\langle M,C\rangle = \mathsf{P}_M(\exists i \leqslant n.S_i \in C)$. We can see that the set $(\forall i \leqslant n.S_i \notin C)$ is the complement set of $(\exists i \leqslant n.S_i \in C)$. Therefore we also know the following:

$$\mathsf{P}_M(\forall i \leqslant n.S_i \notin C) = 0 \tag{4.101}$$

Now, let us assume that there exists a non-terminating program run $t \in \overline{T}(M)$. Due to the basic Markov chain properties, i.e., Theorem 2.18, due to Def. 2.16 and due to $t_0 = M$ we know

$$\mathsf{P}_M(S_0 = t_0 \cap \ldots \cap S_n = t_n) = \prod_{i=0}^{n-1} (t_i \to t_{i+1}) \tag{4.102}$$

Now, we know that $t_i \to t_{i+1} > 0$ for all $i \in \mathbb{N}_0$, because t is a program run. Therefore we know

$$\mathsf{P}_M(S_0 = t_0 \cap \ldots \cap S_n = t_n) > 0 \tag{4.103}$$

We know that $t_i \notin C$ for all $i \in \mathbb{N}_0$, because t is a non-terminating program run. Therefore we know that $(S_0 = t_0 \cap \ldots \cap S_n = t_n) \subseteq (\forall i \leqslant n.S_i \notin C)$. Therefore we know that $\mathsf{P}_M(\forall i \leqslant n.S_i \notin C) \geqslant \mathsf{P}_M(S_0 = t_0 \cap \ldots \cap S_n = t_n)$. Therefore, and due to Eqn. (4.103), we know that $\mathsf{P}_M(\forall i \leqslant n.S_i \notin C) > 0$ which contradicts Eqn. (4.101) and therefore completes the proof. □

5

Denotational Semantics

In this chapter we define a denotational semantics for the probabilistic lambda calculus. We will investigate the correspondence between this denotational semantics and the Markov chain semantics.

Distinct denotational semantics differ from each other in the domains they choose as their basis. The choice of mathematical structures depends on the concrete phenomena that the denotational semantics is intended to examine and model. Also, the targeted level of abstraction influences the choice of domains. The original Scott-Strachey approach was based on complete lattices [243, 253]. A common choice for the domain is a directed complete partial order [117, 118]. In particular, also the denotational semantics given for PCF in [223] is based on directed complete partial orders. Another choice often found is an ω-cpo. They are used as domains, e.g., by [192, 224, 117]. In particular, if computability is the subject of interest, i.e., if modeling of generalized notions of computability is the target, ω-cpos enter the stage. Then, they come as ω-algebraic cpos, see [224]. Also, our denotational semantics is based on ω-cpos. For our purposes, they are the ideal basis for denotational semantics. This is so, because our base domains are probability (pre-)distributions, see Def 5.1, which, of course, have the real numbers as their target domain. Now, ω-chains of real numbers are, by definition, monotonically increasing sequences of real numbers. As such, they are subject to the monotone convergence Theorem of real numbers, which therefore bridges between denotational semantics and the theory of real numbers.

In Sect. 5.1 we define the denotational semantics of the probabilistic lambda calculus, by specifying its domains and giving its semantic equations. Sect. 5.2 is a technical section that establishes the well-definedness of the denotational semantics, and provides further semantical tools needed for the analysis of the denotational semantics. Then, Sect. 5.3 proves the basic correspondence between the denotational and the operational semantics. In Sect. 5.4 we compile a list of important work in the field of denotational semantics of probabilistic programming languages. It is rather an individual snapshot of the field. The purpose of Sect. 5.4 is to give the reader an impression of what has been achieved in the field. The section overviews groundbreaking work in the field, work that is important due to its theoretical insight, technical maturity, recentness of results etc. As such, the section cannot be complete. The purpose is to provide the reader entry points into the impressive body of knowledge concerning domains for higher-type probabilistic computing. Probabilistic powerdomains will be a dominating topic in this section.

5.1 Domains and Denotations

5.1.1 Semantics of Types

Given a program, the evaluation semantics assigns a probability to each constant, i.e., the probability with which the program reduces to that constant – see Def 3.6. The constants are considered the result values of the program. However, in general, the evaluation semantic function does not form a probability distribution, i.e., the reachability probabilities of all constants do not sum up to one in general. This is exactly what the termination degree was about. In case the termination degree is one, all constant probabilities sum up to one and the evaluation semantics is a probability distribution. Obviously, we know that the termination degree can never exceed a value of one. In general the sum of probabilities of terms reachable by a program can, of course, exceed the value of one. For example the program $M = \mu\lambda x. +1(x)$ reaches all terms of the form $+1^n(M)$ with probability one, for all numbers $n \geqslant 1$.

With our denotational semantics we want to give a semantics that mimics the evaluation semantics as yielded by our operational semantics. Therefore, the semantics of a program turns out to be what we call a pre-distribution, which is a function with values in the interval $[0,1]$ that sum up to less than or equal to one.

Definition 5.1 (Pre-Distribution) *A function $d : S \longrightarrow [0,1]$ is called a discrete probability pre-distribution, or pre-distribution for short* **iff** *its domain S is countably infinite and its total mass is less than or equal to one, i.e.,*

$$\sum_{s \in S} d(s) \leqslant 1 \tag{5.1}$$

Pre-distributions are used as semantics for programs that have a termination degree less than one. Besides that they are also needed in case of programs with termination degree one. Here, in cases where the program contains a μ-recursion, they are needed as semantics of the finite approximation of the program semantics. Once we have understood that the semantics of programs are, in general, pre-distributions, we can call them also just distributions on some informal occasions.

We denote the set of truth values, as usual, by $\mathbb{B}$. We use $\mathcal{T}$ and $\mathcal{F}$ to denote the respective semantic truth values. We do so, in order to distinguish them from the lambda constants $\dot{t}$ and $\dot{f}$ in semantic equations. We use the usual number figures $0, 1, \ldots$ to denote the elements of the set of natural numbers $\mathbb{N}_0$. There is no risk of confusing them with the constant symbols n_i for natural numbers, because we use the usual number figures only in informal examples, but not in semantic equations.

Next, we need to see that the interval $[0,1]$ with its natural ordering and zero as bottom element forms an ω-cpo. We will use $[0,1]_\omega$ to denote $[0,1]$ in its role as an ω-cpo. We say that $[0,1]_\omega$ is our domain of probability values.

Definition 5.2 (Domain of Probability Values) *We define the domain of probability values, denoted by* $[0,1]_\omega$, *as the* ω*-cpo* $([0,1], \sqsubseteq, \bot)$ *for all* $i, j \in [0,1]$ *as follows:*

$$i \sqsubseteq j \Leftrightarrow i \leqslant j$$
$$\bot = 0$$

The definition of the domain of probability values contains an implicit proposition, i.e., the fact that $[0,1]_\omega$ actually forms an ω-cpo as defined in Def. 2.62. First, as a matter of course, zero is the least element in $[0,1]_\omega$. Therefore, to show that Def. 5.2 is well defined it remains to show that each ω-chain in $[0,1]_\omega$ has a least upper bound in $[0,1]_\omega$. But this is easy. It immediately follows from the least-upper-bound property of real numbers; see Axiom 2.88. The least-upper-bound property is a fundamental property of the real numbers which states that every non-empty set of real numbers that has an upper bound also has a least upper bound. Therefore, with 1 as the bound for all numbers of the interval $[0,1]_\omega$ we have that each ω-chain $i_0 \sqsubseteq i_1 \sqsubseteq i_2 \cdots$ in $[0,1]_\omega$ has a least upper bound. Note that we even do not need the fact that the set $\{i_0, i_1, i_2 \ldots\}$ forms a chain. It suffices that the set has 1 as an upper bound.

Now, the semantics of types is given recursively for all types $t_1, t_2 \in T$ by the following ω-cpos:

$$[\![num]\!] = \lfloor\, \mathbb{N}_0 \longrightarrow [0,1]_\omega \,\rfloor \tag{5.2}$$

$$[\![bool]\!] = \lfloor\, \mathbb{B} \longrightarrow [0,1]_\omega \,\rfloor \tag{5.3}$$

$$[\![t_1 \rightarrow t_2]\!] = [[\![t_1]\!] \longrightarrow [\![t_2]\!]] \tag{5.4}$$

Eqns. (5.2) and (5.3) rely on the definition of function spaces that have an ω-cpo as a target domain; compare with Def. 2.63. Next, Eqn. (5.4) relies on the definition of ω-continuous function spaces in Def. 2.72. The following equations do not provide different information, but show the semantic domains more explicitly for more easy reference later:

$$[\![num]\!] = (\mathbb{N}_0 \rightarrow [0,1]_\omega,\ \sqsubseteq_{num},\ \bot_{num}) \tag{5.5}$$

$$d_1 \sqsubseteq_{num} d_2 \Leftrightarrow \forall n \in \mathbb{N}_0 \,.\, d_1(n) \leqslant d_2(n) \tag{5.6}$$
$$\bot_{num} = n \in \mathbb{N}_0 \mapsto 0 \tag{5.7}$$

$$[\![bool]\!] = \left(\mathbb{B} \to [0,1]_\omega,\ \sqsubseteq_{bool},\ \bot_{bool}\right) \tag{5.8}$$
$$d_1 \sqsubseteq_{bool} d_2 \Leftrightarrow \forall b \in \mathbb{B} \,.\, d_1(b) \leqslant d_2(b) \tag{5.9}$$
$$\bot_{bool} = b \in \mathbb{B}_0 \mapsto 0 \tag{5.10}$$

$$[\![t_1 \to t_2]\!] = \left(\{f : [\![t_1]\!] \to [\![t_2]\!] \mid f \textit{ is } \omega\textit{-continuous}\},\ \sqsubseteq_{t_1 \to t_2},\ \bot_{t_1 \to t_2}\right) \tag{5.11}$$
$$f_1 \sqsubseteq_{t_1 \to t_2} f_2 \Leftrightarrow \forall x \in [\![t_1]\!] \,.\, f_1(x) \sqsubseteq_{t_2} f_2(x) \tag{5.12}$$
$$\bot_{t_1 \to t_2} = x \in [\![t_1]\!] \mapsto \bot_{t_2} \tag{5.13}$$

We define the set $[\![T]\!]$ of all domains and the set $(\!|T|\!)$ of all domain objects as follows:

$$[\![T]\!] = \{[\![t]\!] \mid t \in T\} \tag{5.14}$$
$$(\!|T|\!) = \cup[\![T]\!] \tag{5.15}$$

Furthermore, we define the set of base domains $[\![T_g]\!]$, i.e., domains for the ground types, and the set of functional domains $[\![T{\to}T]\!]$, which is the set of non-base domains. Similarly, we also define the set $(\!|T_g|\!)$ of all base objects and the set of all functionals $(\!|T{\to}T|\!)$ as follows:

$$[\![T_g]\!] = \{[\![num]\!], [\![bool]\!]\} \tag{5.16}$$
$$[\![T{\to}T]\!] = [\![T]\!] \backslash [\![T_g]\!] \tag{5.17}$$

$$(\!|T_g|\!) = [\![num]\!] \cup [\![bool]\!] \tag{5.18}$$
$$(\!|T{\to}T|\!) = (\!|T|\!) \backslash (\!|T_g|\!) \tag{5.19}$$

In Lemma 3.1 we have expressed that each type has a right-associate top-level structure. Of course, a respective form also exists for domains, as expressed by Lemma 5.3. Again this top-level functional structure of domains can be useful as a basis for natural induction proofs.

Lemma 5.3 (Top-Level Functional Structure of Domains) *Each domain* $D \in (\!|T|\!)$ *has the form* $[D_n \to [\ldots[D_2 \to [D_1 \to B]]...]]$ *for some* $n \geqslant 0$ *with* $D_i \in (\!|T \to T|\!)$ *and a base domain* $B \in [\![T_g]\!]$.

Proof. Corollary from Lemma 3.1 □

5.1.2 Semantics of Lambda Terms

The crucial aspect of denotational semantics is that it strictly follows the principle of decomposition. A denotational semantics assigns the same kind of mathematical object to each term of a language, a domain object from $(\!| T |\!)$ in our case. It does so along the abstract syntax of the respective language. Each term of a language is composed from sub-terms by the application of a syntactical constructor; compare also with Def. 3.2. A denotational semantics follows this composition of terms. It assigns a semantics to a term by recursively applying a semantic function to its sub-terms. In the case of lambda calculi the decomposition principle of denotational semantics implies that we also need to specify mathematical objects for free variables and open terms. Therefore, we need a concept of variable environment which fixes a mathematical object for each free variable. Closed terms are independent of variable environments, however open terms receive a unique semantics only via a given variable environment.

Variable Environments

The set of variable environments $\mathcal{E}$, also environments for short, is defined as the set of type-respecting functions from variables to domains. The undefined environment $\emptyset : \mathcal{E}$, also called the empty environment, is defined as the environment that assigns to each variable the bottom element of the corresponding domain.

Definition 5.4 (Variable Enviroments) *The set $\mathcal{E}$ of variable environments, also environments for short, is defined as follows:*

$$\mathcal{E} = \{ \rho : Var \rightarrow (\!| T |\!) \mid \forall x_t \in Var \,.\, \rho(x_t) \in [\![t]\!] \} \tag{5.20}$$

Definition 5.5 (Empty Environment) *The empty environment $\emptyset : \mathcal{E}$ is defined as follows:*

$$\emptyset \;=\; x_t \in Var \mapsto \bot_t \tag{5.21}$$

We have not defined environments as families of functions indexed by types, but as type-respecting functions from the set of all variables into the set of all domain objects. This way, we save effort in maintaining type information. Given an environment $\rho : \mathcal{E}$ we can simple write $\rho(x_t)$ and do not have to care about the type t of the variable x_t.

Semantic Equations

The denotational semantics assigns a domain object to each lambda term, with respect to a given variable environment. As usual, we use semantic bracketing as notation for the denotational semantics, i.e., we denote the semantics of a term M with respect to an environment ρ by $[\![M]\!]_\rho$. We can make the semantic function more explicit as follows:

$$[\![\underline{\ \ }_2]\!]_{\underline{\ \ }_1} : \mathcal{E} \longrightarrow (\Lambda \longrightarrow (\![T]\!)) \tag{5.22}$$

Note that the semantic function as specified in Eqn. (5.22) will be again a type-respecting function for each environment ρ, i.e., if a lambda term $M : t$ has type t then its semantics $[\![M]\!]_\rho$ is a domain object in $[\![t]\!]$. However, this property of type preservation is not immediately given. Rather we must show that it holds. Actually, this property is exactly what the well-definedness in Sect. 5.2 is about. Basically, the well-definedness of the semantic specification function is about the question whether all of the defined semantic objects actually exist. Here, it is crucial that all of the semantic objects have certain properties, in particular, it is necessary that all of the involved functions at higher types must be ω-continuous. This is necessary because the definition of recursion relies on the semantic fixed-point operator, which relies on ω-continuous functions. Now, we have defined our domains as structures of objects that show this important property. Therefore, the question of well-definedness can be grasped as the question of semantic type preservation, as we will see in Sect. 5.2. For each given environment $\rho : \mathcal{E}$, we define the denotational semantics for all terms $M, N, N_1, N_2 \in \Lambda$ by a series of the following equations, Eqns. (5.23) through (5.34).

Some of the semantic equations rely on the existence of basic semantic operations $[\![+1]\!], [\![-1]\!]$ and $[\![0?]\!]$, which are defined in due course in Sect. 5.1.3. Furthermore, the semantic equations rely on a functional addition operator $M \oplus N$ and a scalar multiplication operator $i \otimes M$ for terms M and probabilities i that are defined in Sect. 5.1.4. The operators are needed to define the semantics of the choice operation $M|N$ as $(0.5 \otimes [\![N_1]\!]_\rho) \oplus (0.5 \otimes [\![N_2]\!]_\rho)$. In case the terms M and N are programs, the effect of this should be intuitively clear. It yields a pointwise weighted average of the two pre-distributions that the terms M and N stand for. In this case the weights are both 0.5 and therefore it actually delivers the plain average of the two respective pre-distributions. With respect to the syntactical choice operator, we are currently only interested in that special case. Nevertheless, the general case is also needed in our semantic equations, i.e., in the specification of the conditional expressions. Here the weights stem from the probabilities that the condition is $\dot{t}$ or $\dot{f}$ and can have any value. For terms of higher type, the definition of the semantic choice propagates the probability weights down to the involved flat domains and this way appropriately merges the semantic objects. We will see how this works in due course in Sect. 5.2.1, where we provide the definitions of the semantic operators $\oplus$ and $\otimes$ and investigate them further. Furthermore, the equations exploit the notation for function update, see Def. 2.84, which is used to denote the update of the variable environment ρ. Furthermore, we need the fixed-point operator Φ as defined in Def. 2.80 to give semantics to recursion, i.e., the μ-operator.

$$\forall\, v \in Var: \; [\![v]\!]_\rho = \rho(v) \tag{5.23}$$

$$\forall n_i \in C_{num}: \; [\![n_i]\!]_\rho = n \in \mathbb{N}_0 \mapsto \begin{cases} 1 & , n = i \\ 0 & , else \end{cases} \tag{5.24}$$

$$[\![\dot{t}]\!]_\rho = b \in \mathbb{B} \mapsto \begin{cases} 1 & , b = \mathcal{T} \\ 0 & , b = \mathcal{F} \end{cases} \tag{5.25}$$

$$[\![\dot{f}]\!]_\rho = b \in \mathbb{B} \mapsto \begin{cases} 0 & , b = \mathcal{T} \\ 1 & , b = \mathcal{F} \end{cases} \tag{5.26}$$

$$[\![+1(M)]\!]_\rho = [\![+1]\!]\, [\![M]\!]_\rho \tag{5.27}$$

$$[\![-1(M)]\!]_\rho = [\![-1]\!]\, [\![M]\!]_\rho \tag{5.28}$$

$$[\![0?(M)]\!]_\rho = [\![0?]\!]\, [\![M]\!]_\rho \tag{5.29}$$

$$[\![\mathit{if}\, M\ \mathit{then}\ N_1\ \mathit{else}\ N_2]\!]_\rho = [\![M]\!]_\rho(\mathcal{T}) \otimes [\![N_1]\!]_\rho \;\oplus\; [\![M]\!]_\rho(\mathcal{F}) \otimes [\![N_2]\!]_\rho \tag{5.30}$$

$$[\![\lambda x_t.M]\!]_\rho = d \in [\![t]\!] \mapsto \left([\![M]\!]_{\rho[x_t:=d]}\right) \tag{5.31}$$

$$[\![MN]\!]_\rho = [\![M]\!]_\rho\, [\![N]\!]_\rho \tag{5.32}$$

$$[\![\mu M]\!]_\rho = \Phi(\, [\![M]\!]_\rho \,) \tag{5.33}$$

$$[\![M|N]\!]_\rho = 0.5 \otimes [\![M]\!]_\rho \;\oplus\; 0.5 \otimes [\![N]\!]_\rho \tag{5.34}$$

5.1.3 Basic Semantic Operators

In this section we define the basic semantic operators of addition, subtraction and test for zero.

Definition 5.6 (Basic Semantic Operators) *We define the functions* $[\![+1]\!] : [\![num]\!] \to [\![num]\!]$, $[\![-1]\!] : [\![num]\!] \to [\![num]\!]$ *and* $[\![0?]\!] : [\![num]\!] \to [\![bool]\!]$ *as follows:*

$$[\![+1]\!] = d \in [\![num]\!] \mapsto \left(n \in \mathbb{N}_0 \mapsto \begin{cases} d(n-1) & , n > 0 \\ 0 & , else \end{cases} \right) \tag{5.35}$$

$$[\![-1]\!] = d \in [\![num]\!] \mapsto \left(n \in \mathbb{N}_0 \mapsto \begin{cases} d(n+1) & , n > 0 \\ d(0) + d(1) & , else \end{cases} \right) \tag{5.36}$$

$$[\![0?]\!] = d \in [\![num]\!] \mapsto \left(b \in \mathbb{B} \mapsto \begin{cases} d(0) & , b = \mathcal{T} \\ \sum\limits_{n \geqslant 1} (d(n)) & , b = \mathcal{F} \end{cases} \right) \tag{5.37}$$

Let us have a look at how the semantics of the test for zero $[\![0?]\!]d$ is defined for a datum $d \in [\![num]\!]$, in particular, at how $([\![0?]\!]\ d)(\mathcal{F})$ is defined as a sum. First, the semantics of $([\![0?]\!]\ d)(\mathcal{T})$ is defined as $d(0)$. However, the semantics of $([\![0?]\!]\ d)(\mathcal{F})$ cannot be defined simply as $1 - ([\![0?]\!]\ d)(\mathcal{T})$. Instead, we must sum up the $d(n)$ of all $n \geqslant 1$ different from zero to receive $([\![0?]\!]\ d)(\mathcal{F})$. This is so, because d is not a distribution but only a pre-distribution, i.e., its total mass may be less than one leaving open room to represent non-termination of a program. The denotational semantics is designed to correspond to the operational semantics. Therefore, given a program M, we have that

$$(M \Rightarrow \dot{t}) + (M \Rightarrow \dot{f}) \leqslant 1 \tag{5.38}$$

$$([\![0?]\!]\ [\![M]\!])(\mathcal{T}) + ([\![0?]\!]\ [\![M]\!])(\mathcal{F}) \leqslant 1 \tag{5.39}$$

The sums in Eqns. (5.38) and (5.39) are less than one exactly if the degree of non-termination $\eta_M\langle S, \overline{C}\rangle$ of M is different from zero.

Based on Def. 5.6 the semantic equations in Eqns. (5.27), (5.28) and (5.29) can be rewritten into a more direct form. For all terms $M \in \Lambda$ we have that

$$[\![+1(M)]\!]_\rho = n \in \mathbb{N}_0 \mapsto \begin{cases} [\![M]\!]_\rho(n-1) & , n > 0 \\ 0 & , else \end{cases} \tag{5.40}$$

$$[\![-1(M)]\!]_\rho = n \in \mathbb{N}_0 \mapsto \begin{cases} [\![M]\!]_\rho(n+1) & , n > 0 \\ [\![M]\!]_\rho(0) + [\![M]\!]_\rho(1) & , else \end{cases} \tag{5.41}$$

$$[\![0?(M)]\!]_\rho = b \in \mathbb{B} \mapsto \begin{cases} [\![M]\!]_\rho(0) & , b = \mathcal{T} \\ \sum\limits_{n \geqslant 1} ([\![M]\!]_\rho(n)) & , b = \mathcal{F} \end{cases} \tag{5.42}$$

It would also be usual to give the semantic equations for $[\![+1(M)]\!]$, $[\![-1(M)]\!]$ and $[\![0?(M)]\!]$ immediately in their direct forms of Eqns. (5.40), (5.41) and (5.42) and, actually, the direct equations are more easy to read, because they work with one abstraction fewer. However, we need to make the basic operators explicit as semantic operators, because we need to investigate them in their own right. In particular, we need to prove that they are ω-continuous; compare with Lemma 5.20.

5.1.4 Higher-Type Arithmetics

The semantics of probabilistic choices $M|N$ and conditional expressions *if* M *then* N_1 *else* N_2 in Eqns. (5.34) and (5.30) rely on the existence of higher-type operators for addition of functionals and scalar multiplication of a number with a functional, which we denote $\oplus$ and $\otimes$ respectively. We will call these operators also functional addition and functional scalar multiplication in the sequel. The operators work for arbitrarily nested functionals as long as the right-most or inner-most target type equals the set of real numbers $\mathbb{R}$, i.e., the operators work for objects that have a type of the following form:

$$(S_n \rightarrow (\ldots(S_2 \rightarrow (S_1 \rightarrow (S_0 \rightarrow \mathbb{R})))...)) \tag{5.43}$$

In accordance with our semantic domains we call the elements of S_0 in Eqn. (5.43) data points. Eventually, the operators are defined in terms of flat operations on real numbers and this way the usual arithmetical laws propagate to the higher-type operators. All the higher-type operators are defined for arbitrary sets as arguments and the full set of real numbers $\mathbb{R}$. In due course, in Sect. 5.2, it will be our task to show that the usage of the higher-type operators in our semantic equations does not lead us out of the allowed domains of ω-continuous functions.

Definition 5.7 (Functional Addition) *Functional addition is defined as a family of operators* $\oplus : S \times S \rightarrow S$ *for all sets of functionals* S *of the form* $S_n \rightarrow \ldots S_1 \rightarrow S_0 \rightarrow \mathbb{R}$, *objects* $f, f' \in S$ *and objects* $e_i \in S_i$ *for all* $i \leqslant n$ *as follows:*

$$(f \oplus f')\, e_n \ldots e_0 = (f\, e_n \ldots e_0) + (f'\, e_n \ldots e_0) \tag{5.44}$$

Definition 5.8 (Functional Scalar Multiplication) *Functional scalar multiplication is defined as a family of operators* $\otimes : \mathbb{R} \times S \rightarrow S$ *for all sets of functionals* S *of the form* $S_n \rightarrow \ldots S_1 \rightarrow S_0 \rightarrow \mathbb{R}$, *real numbers* $k \in \mathbb{R}$, *objects* $f \in S$ *and objects* $e_i \in S_i$ *for all* $i \leqslant n$ *as follows:*

$$(k \otimes f)\, e_n \ldots e_0 = k \cdot (f\, e_n \ldots e_0) \tag{5.45}$$

Definition 5.9 (Functional Null Element) *For each set of functionals* S *of the form* $S_n \rightarrow \ldots S_1 \rightarrow S_0 \rightarrow \mathbb{R}$ *we define the functional null element* 0_S *such that for all objects* $e_i \in S_i$ *for all* $i \leqslant n$ *the following holds:*

$$0_S\, e_n \ldots e_0 = 0 \tag{5.46}$$

Lemma 5.10 (Functional Operator Vector Space) *For each set of functionals* S *of the form* $S_n \rightarrow \ldots S_1 \rightarrow S_0 \rightarrow \mathbb{R}$ *the algebraic structure* $(S, \oplus, \otimes, 0_S)$ *forms a vector space.*

Corollary 5.11 (Functional Null Elements of Domains) *Given a domain* $(D, \sqsubseteq_D, \bot_D)$ *in* $[\![T]\!]$, *we have that* $\bot_D = 0_D$.

Definition 5.12 (Functional Summation) *Functional summation is defined as a family of operators* $\sum : (\omega \to S) \to S$ *for all sets of functionals* S *of the form* $S_n \to \ldots S_1 \to S_0 \to \mathbb{R}$, *countable sets of objects* $(f_i \in S)_{i\in\omega}$ *and objects* $e_j \in S_j$ *for all* $j \leqslant n$ *as follows:*

$$\left(\sum_{i=0}^{\infty} f_i\right) e_n \ldots e_0 = \sum_{i=0}^{\infty}(f_i\, e_n \ldots e_0) \tag{5.47}$$

Lemma 5.13 (Applications Distribute over Functional Operators) *Given objects of some higher type* $f, f' \in S \to S'$, *a real number* $i \in \mathbb{R}$ *and an object* $e \in S$ *we have that*

$$(f \oplus f')\, e = (fe) \oplus (f'e) \tag{5.48}$$

$$i \otimes (fe) = (i \otimes f)e \tag{5.49}$$

$$\left(\sum_{f\in F} f\right)e = \sum_{f\in F}(f(e)) \tag{5.50}$$

Lemma 5.14 (ω-Continuity of Functional Operators) *Given an* ω*-cpo* $D = [D_n \to [\ldots [D_2 \to [D_1 \to \mathbb{R}]]...]]$, ω*-chains* $(f_i \in D)_{i\in\omega}$ *and* $(f'_i \in D)_{i\in\omega}$, *a sequence of real numbers* $(i_n)_{n\in\omega}$ *and a set of* ω*-chains* $F \subseteq \omega \to D$ *we have that*

$$\bigsqcup_{n\in\omega} f_n \oplus \bigsqcup_{n\in\omega} f'_n = \bigsqcup_{n\in\omega} (f_n \oplus f'_n) \tag{5.51}$$

$$\lim_{n\to\infty} i_n \otimes \bigsqcup_{m\in\omega} f_m = \bigsqcup_{k\in\omega} (i_k \otimes f_k) \tag{5.52}$$

$$\sum_{f\in F}(\bigsqcup_{n\in\omega} f_n) = \bigsqcup_{n\in\omega}\left(\sum_{f\in F} f_n\right) \tag{5.53}$$

5.1.5 Further Remarks on Domains and Denotations

A natural way to deal with non-determinism in programming languages is to introduce sets of objects as denotations for programs. This gives rise to powerdomain structures in the literature. Several kinds of powerdomains have been studied [221, 222, 246, 247]. Essentially, the several options differ in the way they treat and represent non-terminating program runs. Our base objects are always single pre-distributions. Whenever several pre-distributions arise non-deterministically, i.e., on occasion of the probabilistic choice construct, they are immediately eliminated and flattened into a single new pre-distribution. This procedure is populated from base elements through to the objects of all higher types. The least-upper-bound elimination Lemma, which we have available for ω-cpos in Lemma 2.75, takes part in this technique. The question of how best to represent non-termination by explicit bottom elements is not an issue in our approach. In our approach, non-termination is implicitly represented by the degree of non-termination. Bottom elements emerge automatically in our base domains, as pre-distributions that assign a zero percent probability to all data points. For the same reason, we also do not need to use generalized notions of real number sequences, such as topological generalizations as Moore-Smith Sequences [200, 186]. We are just fine with ordinary convergent sequences of real numbers.

5.2 Well-Definedness of the Denotational Semantics

In this section we prove that the semantic function provided in Sect. 5.1 is actually well defined. Given a term M and an environment ρ the semantic function defines a semantic object by the semantic equations Eqns. (5.23) through (5.34). It remains to show that the defined objects actually exist and actually exist in the correct domains, which is mutually dependent. First, we need to show that all terms of ground type have a pre-distribution as semantics. Furthermore, we need to show, inductively, that the semantic objects belong to the appropriate domains for all terms. This is exactly what Lemma 5.22 is about. This type preservation is quite easy to see for constants and the corresponding basic semantic objects in the domains $[\![num]\!]$ and $[\![bool]\!]$. However, it is not so immediately given for the other, complex constructs of our calculus. For example, we need to argue that the fixed point used to define the recursion actually exists. In order to have that, we need to show that all involved functions are ω-continuous.

Actually, we need to show for all higher-type constructs that the corresponding semantic objects are ω-continuous functions. This necessity is immediately clear for terms of lambda abstractions $\lambda v.M$. Here, given $v : t$ and $M : t'$ the corresponding semantic equation Eqn. (5.31) defines a function in $[\![t]\!] \rightarrow [\![t']\!]$ as the corresponding semantic object. However, the semantic equation itself does not ensure that this function belongs to the required domain $[[\![t]\!] \rightarrow [\![t']\!]]$ of ω-continuous functions, which remains to be proven. Actually, the case of lambda abstraction can be considered as particularly tricky, because it entails a further structural induction on terms of the bodies M of lambda abstractions $\lambda.M$, which shows explicitly in the need for the l.u.b. substitution Lemma 5.21. However, conceptually the case of lambda abstraction is not special. The ω-continuity of corresponding functions has to be proven for all terms of higher type. The proof of this property always follows a similar pattern. In each case, it relies, in one way or the other, on respective properties of corresponding semantic operators. For example, in case of recursion, this is the fixed-point operator Φ that we have investigated in Sect. 2.4.8. In case of the probabilistic choice it is the so-called semantic choice operator that we will elaborate together with the necessary lemmas in the upcoming Sect. 5.2.1.

5.2.1 The Semantic Choice Operator

The semantics of probabilistic choice $M|N$ in Eqn. (5.34) and conditional expressions *if* M *then* N_1 *else* N_2 in Eqn. (5.30) can be expressed succinctly in terms of a semantic choice operator $d \parallel_j^i d'$ for each domain $D \in [\![T]\!]$; see Def. 5.15. Here, i and j are probabilities and the task of the semantic choice operator is to merge together data points d and d' according to their respective probability, and weighted according to the probabilities specified by i and j.

Definition 5.15 (Semantic Choice Operator) *The semantic choice operator is a family of functions* $\parallel_j^i : D \times D \rightarrow D'$ *that is defined for each domain* $D \in [\![T]\!]$, *semantic objects* $f, g \in D$ *and indices* $i, j \in [0,1]$ *as follows:*

$$(f \parallel_j^i g) = (i \otimes f) \ \oplus \ (j \otimes g) \tag{5.54}$$

The semantic choice operator works for all kinds of domains. If it is applied to data from the base domain, its arguments f and g are pre-distributions. In this case, it takes each data point d and yields the weighted average probability it has under the pre-distributions f and g weighted by i and j. This is an arithmetic function. Otherwise, if the arguments of the semantic choice operator are functionals, the choice operator recursively propagates the weights down to data objects of basic domains, as expressed by Corollary 5.16.

Corollary 5.16 (Recursivity of the Semantic Choice Operator) *For all domains* $D \in [\![T]\!]$, *semantic objects* $f, g \in D$ *and indices* $i, j \in [0,1]$ *we have the following:*

$$(f \parallel_j^i g) = \begin{cases} d \in S \mapsto (\ i \cdot f(d) \ + \ j \cdot g(d)\ ) \in [0,1] \ , & f, g \in \lfloor S \rightarrow [0,1]_\omega \rfloor \\ d \in D_1 \mapsto (\ f(d) \ \parallel_j^i g(d)\ ) \in D_2 & , \ f, g \in [D_1 \rightarrow D_2] \end{cases} \tag{5.55}$$

The set S in Eqn. (5.55) can be either $\mathbb{N}_0$ or $\mathbb{B}$. In the case of base domains, the semantic choice operator gets some base objects and also yields a base object, i.e., in this case we know that the semantic choice operator is a function $D \times D \rightarrow D$, where D can be either $\lfloor \mathbb{N}_0 \rightarrow [0,1]_\omega \rfloor$ or $\lfloor \mathbb{B} \rightarrow [0,1]_\omega \rfloor$. This means that in the case of base domains we know that $D' = D$ already by Corollary 5.16. This is different in case of functional domains. Here we know that the arguments f and g are elements of a domain $[D_1 \rightarrow D_2]$. However, all we know about the result $f(d) \parallel_j^i g(d)$ is that it is an element of $D_1 \rightarrow D_2$. It remains to show that $D' = D$ also in this case. As a consequence, we need to show that $f(d) \parallel_j^i g(d)$ preserves ω-continuity. We will do this in Lemma 5.18 and summarize the result that $\parallel_j^i : D \times D \rightarrow D$ for all domains D in Corollary 5.19. Even more, we will have that $\parallel_j^i : [[D \times D] \rightarrow D]$ in Corollary 5.19.

It is important to have this result as a tool to prove the well-definedness of the denotational semantics. As a first step, we will show that the semantic choice operator is itself continuous. For each domain D, we have defined the semantic choice operator as a function in $D \times D \rightarrow D'$. The target is to show that the semantic choice operator belongs to $[[D \times D] \rightarrow D]$.

Lemma 5.17 (ω-Continuity of the Semantic Choice) *The semantic choice operator is ω-continuous in both of its arguments.*

Proof. We need to show that for all ω-chains $C = (c_i)_{i\in\omega}$ and $C' = (c'_i)_{i\in\omega}$ we have that

$$\sqcup C \parallel^i_j \sqcup C' \;=\; \bigsqcup_{n\in\omega} (c_n \parallel^i_j c'_n) \tag{5.56}$$

Due to Lemma 5.3 we can assume that the domains of C and C' have the following form for some $n \geqslant 0$, some possibly empty sequence of domains D_i for all $1 \leqslant i \leqslant n$ and some base domain $B \in [\![T_g]\!]$ of the form $\lfloor S \rightarrow [0,1]_\omega \rfloor$:

$$[D_n \rightarrow [\ldots [D_2 \rightarrow [D_1 \rightarrow B]]...]] \tag{5.57}$$

We show Eqn. (5.56) for all domains by natural induction over the length n of the domain in Eqn. (5.57).

***In Case of n=0*:** In case $n = 0$ we know that the domains of C and C' have the form $\lfloor S \rightarrow [0,1]_\omega \rfloor$ for some S. We show Eqn. (5.56) by proving that $\sqcup C \parallel^i_j \sqcup C'$ equals $\bigsqcup_{n\in\omega} (c_n \parallel^i_j c'_n)$ for all data points $p \in S$. Given an arbitrary data point p we start with the following:

$$(\sqcup C \parallel^i_j \sqcup C')\, p \tag{5.58}$$

Now, Corollary 5.16 applies. We are concerned with the first clause of Eqn. (5.55), which deals with the case of data points. Due to Corollary 5.16 we know that Eqn. (5.58) equals

$$(i \cdot ((\sqcup C)p) + (j \cdot (\sqcup C')p)) \tag{5.59}$$

Due to the pointwise construction of least upper bounds of ω-cpo targeting functions, i.e., Lemma 2.64, we know that Eqn. (5.59) equals

$$(i \cdot \bigsqcup_{n\in\omega} c_n(p)) + (j \cdot \bigsqcup_{n\in\omega} c'_n(p)) \tag{5.60}$$

We know that the ω-chains $(c_n(p))$ and $(c'_n(p))$ in Eqn. (5.60) are monotonically increasing sequences of real numbers; compare with Def. 2.92. Furthermore, we have that all sequences in $[0,1]$ are necessarily bounded from above, with 1 as absolute bound. Therefore the monotone convergence Theorem 2.93 applies, which ensures that the least upper bounds of these sequences equal the limits of these sequences in the sense of convergence in real analysis. Therefore, altogether we have that Eqn. (5.60) is equivalent to the following formula:

$$(i \cdot \mathit{lim}_{n\rightarrow\infty}\, c_n(p)) + (j \cdot \mathit{lim}_{i\rightarrow\infty}\, c'_n(p)) \tag{5.61}$$

Due to the limit properties of sequences of real numbers, see Lemma 2.91, we know that Eqn. (5.61) equals

$$\lim_{n\to\infty} (i \cdot c_n(p) + j \cdot c'_n(p)) = \bigsqcup_{n\in\omega} (i \cdot c_n(p) + j \cdot c'_n(p)) \tag{5.62}$$

Again, due to Corollary 5.16 we know that Eqn. (5.62) equals

$$\bigsqcup_{n\in\omega} (\,(c_n \parallel^i_j c'_n)\, p) \tag{5.63}$$

Finally, again due to the pointwise construction of least upper bounds of functions in Lemma 2.64 we know that Eqn. (5.63) equals

$$(\,\bigsqcup_{n\in\omega} (c_n \parallel^i_j c'_n)\,)\, p \tag{5.64}$$

In Case of $n > 0$**:** In case $n > 0$ we have that C and C' have the form $[D_{n+1} \to [D_n \to [\ldots [D_2 \to [D_1 \to B]]...]]]$. We show Eqn. (5.56) by proving that $\sqcup C \parallel^i_j \sqcup C'$ equals $\bigsqcup_{n\in\omega} (c_n \parallel^i_j c'_n)$ for all objects $d \in D_{n+1}$. Given an arbitrary object $d \in D_{n+1}$ we start with the following:

$$(\sqcup C \parallel^i_j \sqcup C')\, d \tag{5.65}$$

Due to Eqn. (5.55), second clause, we know that Eqn. (5.65) equals

$$(\sqcup C) d \;\parallel^i_j\; (\sqcup C')\, d \tag{5.66}$$

Due to the construction of least upper bounds of functions in Lemma 2.64 we know that Eqn. (5.66) equals

$$\bigsqcup_{n\in\omega} (c_n(d)) \;\parallel^i_j\; \bigsqcup_{n\in\omega} (c'_n(d)) \tag{5.67}$$

Let us consider the ω-chains $(c_n(d))_{n\in\omega}$ and $(c'_n(d))_{n\in\omega}$. In these, the functions $c_n(d)$ and $c'_n(d)$ are elements of the domain $[D_n \to [\ldots [D_2 \to [D_1 \to B]]...]]$. Therefore, the induction hypothesis applies and we have that Eqn. (5.67) equals

$$\bigsqcup_{n\in\omega} (\,(c_n \parallel^i_j c'_n)\, d\,) \tag{5.68}$$

Now, according to the construction of least upper bounds of functions in Lemma 2.64 we know that Eqn. (5.68) equals

$$(\,\bigsqcup_{n\in\omega} (c_n \parallel^i_j c'_n)\,)\, d \tag{5.69}$$

□

Now, given that the semantic choice operator is ω-continuous we have, as a corollary, that it preserves ω-continuity.

Lemma 5.18 (The Semantic Choice Preserves ω-Continuity) *The semantic choice operator preserves ω-continuity, i.e., for all functional domains $[D \rightarrow E]$, functions $f, f' \in [D \rightarrow E]$ and indices i, j it holds that $(f \parallel^i_j f')$ is an ω-continuous function, i.e., $(f \parallel^i_j f') \in [D \rightarrow E]$.*

Proof. This Lemma follows as a corollary from Lemma 5.17. We need to show that for all ω-chains $D' = (d_n)_{n\in\omega}$ we have that

$$(f \parallel^i_j f')(\sqcup D') \;=\; \bigsqcup_{n\in\omega}(\,(f \parallel^i_j f')\, d_n\,) \tag{5.70}$$

By Corollary 5.16 we know that $(f \parallel^i_j f')(\sqcup D')$ equals

$$f(\sqcup D') \parallel^i_j f'(\sqcup D') \tag{5.71}$$

Now, we know that both f and f' are ω-continuous. Therefore, Eqn. (5.71) equals

$$\sqcup(f(d_n))_{i\in\omega} \;\parallel^i_j\; \sqcup(f'(d_n))_{i\in\omega} \tag{5.72}$$

Now, we have that Lemma 5.17 applies, i.e., we know that the semantic choice operator itself is ω-continuous in both of its arguments. Therefore, we know that Eqn. (5.72) equals

$$\bigsqcup_{n\in\omega}(\,f(d_n) \parallel^i_j f'(d_n)\,) \tag{5.73}$$

Again, by Corollary 5.16 we have that Eqn. (5.73) equals

$$\bigsqcup_{n\in\omega}(\,(f \parallel^i_j f')\, d_n\,) \tag{5.74}$$

□

Next, we will summarize the results of Lemma 5.17 and Lemma 5.18 into a single corollary. Both aspects of the corollary, i.e., ω-continuity as well as the preservation of ω-continuity, are important for the well-definedness of the denotational semantics.

Corollary 5.19 (Retraction of the Semantic Choice Operator) *The semantic choice operator $\parallel^i_j$ is an element of $[[D \times D] \rightarrow D]$ for all domains D.*

Proof. Corollary from Lemma 5.17 and Lemma 5.18 □

5.2.2 Well-Definedness of the Basic Semantic Operators

Each term consists of a constructor and component terms; see also Def. 3.2. Now, it is a general pattern that the semantics of a term is given as the application of a constructor-specific semantic operator to the semantics of the component terms. This is so in case of the probabilistic choice and the conditional expressions that both exploit the choice operator $\parallel^i_j$ as semantic operator. In case of recursion the fixed point Φ serves as semantics operator.

In case of application the operator *apply* introduced in Def. 2.73 can be considered as the corresponding semantic operator. The only case that slightly deviates from this pattern is the case of lambda-abstracting terms. With the basic semantic functions from Def. 5.6 also the semantic definitions of $+1(M)$, $-1(M)$ and $0?(M)$ follow the described pattern of semantic description.

Lemma 5.20 (ω-Continuity of Basic Semantic Functions) *The semantic functions $[\![+1]\!]$, $[\![-1]\!]$ and $[\![0?]\!]$ are ω-continuous functions, i.e., for $C : \omega \to [\![num]\!]$ we have that*

$$[\![+1]\!](\sqcup_{num} C) = \sqcup_{num}([\![+1]\!]^{\dagger} C) \tag{5.75}$$

$$[\![-1]\!](\sqcup_{num} C) = \sqcup_{num}([\![-1]\!]^{\dagger} C) \tag{5.76}$$

$$[\![0?]\!](\sqcup_{num} C) = \sqcup_{bool}([\![zero]\!]^{\dagger} C) \tag{5.77}$$

Proof. We only show that $[\![+1]\!]$ is ω-continuous. The proofs that $[\![-1]\!]$ and $[\![0?]\!]$ are ω-continuous are similar. With $C = (c_i)_{i\in\omega}$ we need to show the following:

$$[\![+1]\!](\sqcup(c_i)_{i\in\omega}) = \sqcup(([\![+1]\!] c_i)_{i\in\omega}) \tag{5.78}$$

We start with the right-hand side of Eqn. (5.78). Due to the definition of $[\![+1]\!]$ in Def. 5.6 we know that the following holds:

$$\sqcup(([\![+1]\!] c_i)_{i\in\omega}) = \bigsqcup \Bigg(\underbrace{n \in \mathbb{N}_0 \mapsto \begin{cases} c_i(n-1) & , n > 0 \\ 0 & , else \end{cases}}_{f_i} \Bigg)_{i\in\omega} \tag{5.79}$$

Now that $(f_i)_{i\in\omega}$ in Eqn. (5.79) is an ω-chain in $[\![num]\!] \to [\![num]\!]$ we can apply Lemma 2.64 so that Eqn. (5.79) equals

$$n \in \mathbb{N}_0 \mapsto \bigsqcup \left(\left(n \in \mathbb{N}_0 \mapsto \begin{cases} c_i(n-1) & , n > 0 \\ 0 & , else \end{cases} \right) n \right)_{i\in\omega} \tag{5.80}$$

After semantic function application we have that Eqn. (5.80) equals

$$n \in \mathbb{N}_0 \mapsto \bigsqcup \left(\begin{cases} c_i(n-1) & , n > 0 \\ 0 & , else \end{cases} \right)_{i\in\omega} \tag{5.81}$$

The conditions in the function definition of Eqn. (5.81), i.e., $n > 0$ or otherwise $n = 0$, are independent of the index $i \in \omega$ of the ω-chain. Therefore we deal with two completely separate cases of ω-chains, i.e., the chain $(c_i(n-1))_{i\in\omega}$ in case $n > 0$ and the chain $(0)_{i\in\omega}$ in case $n = 0$. Therefore, Eqn. (5.81) can be rewritten as follows:

$$n \in \mathbb{N}_0 \mapsto \begin{cases} \sqcup(c_i(n-1))_{i\in\omega} & , n > 0 \\ \sqcup(0)_{i\in\omega} & , else \end{cases} \tag{5.82}$$

With $\bigsqcup(0)_{i\in\omega} = 0$ and the introduction of a further semantic abstraction we can turn Eqn. (5.82) into the following equivalent formula:

$$n \in \mathbb{N}_0 \mapsto \begin{cases} (n \mapsto (\sqcup(c_i(n)))_{i\in\omega})(n-1) & , n > 0 \\ 0 & , else \end{cases} \tag{5.83}$$

Now we have that $(c_i)_{i\in\omega}$ is an ω-chain in $[\![num]\!]$. We know that $[\![num]\!]$ equals $\lfloor \mathbb{N}_0 \rightarrow [0,1]_\omega \rfloor$; see Eqn. (5.2). Therefore, we can apply Lemma 2.64 so that we have that Eqn. (5.83) equals

$$n \in \mathbb{N}_0 \mapsto \begin{cases} (\sqcup(c_i)_{i\in\omega})(n-1) & , n > 0 \\ 0 & , else \end{cases} \tag{5.84}$$

Finally, with Def. 5.6 we have that Eqn. (5.84) equals

$$[\![+1]\!](\sqcup(c_i)_{i\in\omega}) \tag{5.85}$$

□

5.2.3 Well-Definedness of Semantic Objects

As a first step, we prove the so-called l.u.b. substitution Lemma 5.21. The l.u.b. substitution Lemma is a kind of ω-continuity result for all terms. It is a quite technical lemma, which becomes necessary to prove that lambda abstraction terms $\lambda v.M$ have a well-defined semantics as part of Lemma 5.22. Albeit the lemma is quite technical it is a central lemma, because it accumulates and concentrates all the necessary results on the involved semantical operators.

Open terms can receive a semantics only in the context of a variable environment. Now, Lemma 5.21 considers two semantic objects. The first is gained as semantics of a term M after updating its environment at one point by the l.u.b. of a given ω-chain C. The second is gained by updating the environment several times with all elements of the chain C and taking the l.u.b. of the ω-chain of the several resulting semantics of M. The Lemma states that both of these semantic objects are equal. You can consider the lemma also as a statement about the ω-continuity of the semantic function itself. The semantic function has the form $\Lambda \times \mathcal{E} \rightarrow (\!|t|\!)$. Now, the lemma is a statement about the continuity of the semantic function in its second argument, albeit a specialized one, because we consider only chains of environments whose elements grow at exactly one point. For the purpose of proving the well-definedness of the semantic function it is sufficient in its given form.

Lemma 5.21 (An L.U.B. Substitution Lemma) *For all terms $M : t$, environments $\rho : \mathcal{E}$, types t', ω-chains $c = (c_i)_{i\in\omega}$ in $[\![t']\!]$ and variables $v : t'$ we have that*

$$[\![M]\!]_{\rho[v:=\sqcup_{[\![t']\!]}c]} = \sqcup_{[\![t]\!]}([\![M]\!]_{\rho[v:=c_i]})_{i\in\omega} \tag{5.86}$$

Proof. We proceed by structural induction over the construction of terms. Variables and constants form the base cases of the induction, whereas the other cases all exploit the induction hypothesis.

***In Case of Variables*:** In case of a variable x we need to consider two cases for $[\![x]\!]_{\rho[v:=\sqcup_{t'}c]}$, i.e., the case $v = x$ and the case $v \neq x$. In case $v = x$ we know due to the definition of the semantic function in Eqn. (5.23), applied twice, that

$$[\![x]\!]_{\rho[x:=\sqcup_{[\![t']\!]}c]} = \sqcup_{[\![t]\!]}((c_i)_{i\in\omega}) = \sqcup_{[\![t]\!]}(([\![x]\!]_{\rho[x:=c_i]})_{i\in\omega}) \tag{5.87}$$

In case $v \neq x$ the update of v in the environment is not relevant for the semantics of x. Now, given an ω-chain that consists of a single data value for all of its elements, we know that the l.u.b. of this ω-chain also equals all of its members. Therefore, and again due to Eqn. (5.23); we know that

$$[\![x]\!]_{\rho[v:=\sqcup_{[\![t']\!]}c]} = [\![x]\!]_{\rho} = \sqcup_{[\![t]\!]}(([\![x]\!]_{\rho})_{i\in\omega}) = \sqcup_{[\![t]\!]}(([\![x]\!]_{\rho[x:=c_i]})_{i\in\omega}) \tag{5.88}$$

***In Case of Constants*:** In case of constants $k \in C$ the environment is not relevant at all for the semantics. The argumentation in this case therefore equals the case of variables with $v \neq x$, compare with Eqn. (5.88), with the respective semantic equations Eqns. (5.24), (5.25) and (5.26):

$$[\![k]\!]_{\rho[v:=\sqcup_{[\![t']\!]}c]} = [\![k]\!]_{\rho} = \sqcup_{[\![t]\!]}(([\![k]\!]_{\rho})_{i\in\omega}) = \sqcup_{[\![t]\!]}(([\![k]\!]_{\rho[x:=c_i]})_{i\in\omega}) \tag{5.89}$$

***In Case of +1, -1 and 0?*:** We show the result for terms of the form $+1(M)$. The cases $-1(M)$ and $0?(M)$ both follow exactly the same pattern. We start with the following term:

$$[\![+1(M)]\!]_{\rho[v:=\sqcup c]} \tag{5.90}$$

Due to the semantic equation Eqn. (5.27) we have that Eqn. (5.90) equals

$$[\![+1]\!][\![(M)]\!]_{\rho[v:=\sqcup c]} \tag{5.91}$$

Now, due to the induction hypothesis we know that Eqn. (5.91) equals

$$[\![+1]\!]\;(\sqcup(([\![M]\!]_{\rho[v:=c_i]})_{i\in\omega})) \tag{5.92}$$

Due to Lemma 5.20 we know that $[\![+1]\!]$ is an ω-continuous function. Therefore we know that Eqn. (5.92) equals

$$\sqcup(([\![+1]\!]\;[\![M]\!]_{\rho[v:=c_i]})_{i\in\omega}) \tag{5.93}$$

Finally, due to the semantic equation Eqn. (5.27) we have that Eqn. (5.93) equals

$$\sqcup(([\![+1(M)]\!]_{\rho[v:=c_i]})_{i\in\omega}) \tag{5.94}$$

***In Case of Probabilistic Choice*:** In a sense, the case of probabilistic choice is the most interesting case. The probabilistic choice is the construct that makes the difference with standard non-probabilistic calculi. Fortunately, with the propositions of Sect. 5.2.1 we are well prepared to prove this case. Due to the definition of the semantic function in Eqn. (5.34) we know that the following equation holds:

$$[\![M|N]\!]_{\rho[v:=\sqcup_{[\![t']\!]}c]} = [\![M]\!]_{\rho[v:=\sqcup_{[\![t']\!]}c]} \parallel^{0.5}_{0.5} [\![N]\!]_{\rho[v:=\sqcup_{[\![t']\!]}c]} \tag{5.95}$$

Due to the induction hypothesis we know that Eqn. (5.95) equals

$$\sqcup_{[\![t]\!]}(([\![M]\!]_{\rho[v:=c_i]})_{i\in\omega}) \parallel^{0.5}_{0.5} \sqcup_{[\![t]\!]} (([\![N]\!]_{\rho[v:=c_i]})_{i\in\omega}) \tag{5.96}$$

Now, due to Lemma 5.17 we know that the semantic probabilistic choice operator is itself ω-continuous. Therefore, we know that Eqn. (5.96) equals

$$\sqcup_{[\![t]\!]}(([\![M]\!]_{\rho[v:=c_i]} \parallel^{0.5}_{0.5} [\![N]\!]_{\rho[v:=c_i]})_{i\in\omega}) \tag{5.97}$$

Finally, again due to the definition of the semantic function in Eqn. (5.34), we know that Eqn. (5.97) equals

$$\sqcup_{[\![t]\!]}(([\![M|N]\!]_{\rho[v:=c_i]})_{i\in\omega}) \tag{5.98}$$

***In Case of Conditional Expressions*:** The semantics of conditional expressions is defined completely in terms of the semantic choice operator so that the proof in this case exactly follows the pattern demonstrated in case of the probabilistic choice.

***In Case of Applications*:** In case of applications $(MN) : t$ we have that $M : s \to t$ and $N : s$ for some type $s : T$. Now we know, due to the definition of the semantic function in Eqn. (5.32), that the following equation holds:

$$[\![MN]\!]_{\rho[v:=\sqcup_{[\![t']\!]}c]} = [\![M]\!]_{\rho[v:=\sqcup_{[\![t']\!]}c]}\ [\![N]\!]_{\rho[v:=\sqcup_{[\![t']\!]}c]} \tag{5.99}$$

Due to the induction hypothesis we know that Eqn. (5.99) equals

$$\sqcup_{[\![s\to t]\!]}([\![M]\!]_{\rho[v:=c_i]})_{i\in\omega}\ (\sqcup_{[\![s]\!]}([\![N]\!]_{\rho[v:=c_j]})_{j\in\omega}) \tag{5.100}$$

Now, due to the pointwise definition of the least upper bound in Lemma 2.64 we know that Eqn. (5.100) equals

$$\sqcup_{[\![t]\!]}(\ [\![M]\!]_{\rho[v:=c_i]}\ (\sqcup_{[\![s]\!]}([\![N]\!]_{\rho[v:=c_j]})_{j\in\omega})\ )_{i\in\omega} \tag{5.101}$$

Now, we know that $[\![s \to t]\!]$ equals $[[\![s]\!] \to [\![t]\!]]$ and therefore all functions in $[\![s \to t]\!]$ are ω-continuous. Therefore, also all of the functions $[\![M]\!]_{\rho[v:=c_i]}$ for all $i \in \omega$ are ω-continuous. Therefore we know that Eqn. (5.101) equals

$$\sqcup_{[\![t]\!]}(\ \sqcup_{[\![t]\!]}(\ [\![M]\!]_{\rho[v:=c_i]}\ [\![N]\!]_{\rho[v:=c_j]}\ )_{j\in\omega}\ )_{i\in\omega} \tag{5.102}$$

Next, we can make the function application in Eqn. (5.102) explicit by introducing the *apply* operator, which we have defined in Def. 2.73, to Eqn. (5.102), which yields

$$\sqcup_{[\![t]\!]}(\ \sqcup_{[\![t]\!]}(\ apply([\![M]\!]_{\rho[v:=c_i]}, [\![N]\!]_{\rho[v:=c_j]})\ )_{j\in\omega}\ )_{i\in\omega} \tag{5.103}$$

Furthermore, we know due to Lemma 2.74 that the *apply* operator is monotone in both of its arguments. Therefore, it possible to apply the l.u.b. elimination Lemma 2.75 to Eqn. (5.103). Therefore we know that Eqn. (5.103) equals

$$\sqcup_{[\![t]\!]}(\ apply([\![M]\!]_{\rho[v:=c_i]}, [\![N]\!]_{\rho[v:=c_i]})\ )_{i\in\omega} \tag{5.104}$$

Now, by eliminating the explicit representation of function application, we can turn Eqn. (5.104) again into

$$\sqcup_{[\![t]\!]}(\ [\![M]\!]_{\rho[v:=c_i]}\ [\![N]\!]_{\rho[v:=c_i]}\ )_{i\in\omega} \tag{5.105}$$

Now, due to the definition of the semantic function in Eqn. (5.32) we know that Eqn. (5.105) equals

$$\sqcup_{[\![t]\!]}(([\![MN]\!]_{\rho[v:=c_i]})_{i\in\omega}) \tag{5.106}$$

In Case of Abstractions: In case of abstractions $\lambda x.M$ we have that $t = s \to s'$ such that $x : s$ and $M : s'$. We know, due to Eqn. (5.31), that the following equation holds:

$$[\![\lambda x.M]\!]_{\rho[v:=\sqcup_{[\![t']\!]}c]} \ = \ d \in [\![s]\!] \mapsto ([\![M]\!]_{\rho[v:=\sqcup_{[\![t']\!]}c,\ x:=d]}) \tag{5.107}$$

Due to the induction hypothesis we know that Eqn. (5.107) equals

$$d \in [\![s]\!] \mapsto \sqcup_{[\![s']\!]}\ ([\![M]\!]_{\rho[v:=c_i,\ x:=d]})_{i\in\omega} \tag{5.108}$$

Now, we can introduce a further semantic abstraction to Eqn. (5.108) yielding the following equivalent formula:

$$d \in [\![s]\!] \mapsto \sqcup_{[\![s']\!]}\ (\ (e \in [\![s]\!] \mapsto [\![M]\!]_{\rho[v:=c_i,\ x:=e]})\ d\ )_{i\in\omega} \tag{5.109}$$

Now, having the function in the form of Eqn. (5.109) we see that the sequence of functions $(e \in [\![s]\!] \mapsto [\![M]\!]_{\rho[v:=c_i,\ x:=e]})_{i\in\omega}$ forms an ω-chain in $[\![s \to s']\!]$. Now, due to the result for least upper bounds of functions in Lemma 2.64 we know that Eqn. (5.109) equals

$$\sqcup_{[\![s\to s']\!]}(e \in [\![s]\!] \mapsto [\![M]\!]_{\rho[v:=c_i,\ x:=e]}\ )_{i\in\omega} \tag{5.110}$$

As the final step, we know due to the definition of the semantic function in Eqn. (5.31) and the fact that $t = s \to s'$ that Eqn. (5.110) equals

$$\sqcup_{[\![t]\!]}(([\![\lambda x.M]\!]_{\rho[v:=c_i]})_{i\in\omega} \tag{5.111}$$

In Case of Recursion: In case of recursion we consider terms of the form $\mu M : t$ with $M : t \to t$. We know that due to Eqn. (5.33) the following equation holds:

$$[\![\mu M]\!]_{\rho[v:=\sqcup_{[\![t']\!]}c]} = \Phi[\![M]\!]_{\rho[v:=\sqcup_{[\![t']\!]}c]} \tag{5.112}$$

Due to the induction hypothesis we know that Eqn. (5.112) equals

$$\Phi(\sqcup_{[\![t\to t]\!]}([\![M]\!]_{\rho[v:=c_i]})_{i\in\omega}) \tag{5.113}$$

Now, due to Lemma 2.82 we know that the fixed-point operator is itself ω-continuous. Therefore we know that Eqn. (5.113) equals

$$\sqcup_{[\![t]\!]}(\Phi[\![M]\!]_{\rho[v:=c_i]})_{i\in\omega} \tag{5.114}$$

Now, again due to Eqn. (5.33) we know that Eqn. (5.114) equals

$$\sqcup_{[\![t]\!]}([\![\mu M]\!]_{\rho[v:=c_i]})_{i\in\omega} \tag{5.115}$$

□

Now, we turn to the well-definedness of the denotational semantics. As we have already explained, the proposition is about the type preservation of the semantic function. In proving the type preservation of the semantic function we prove the existence of the semantic objects for all terms on the fly. The proof needs to be conducted for all terms by structural induction over the construction of terms. The definition of the type system of the calculus in Sect. 3.1.2 with its typing rules forms the background against which the proposition has to be understood and its proof has to be conducted.

Lemma 5.22 (Well-Definedness of Denotations) *For each term M of type t, and environment $\rho \in \mathcal{E}$ we have that its semantics $[\![M]\!]_\rho$ is an object in the semantics of its corresponding domain $[\![t]\!]$, i.e.,*

$$M : t \implies [\![M]\!]_\rho \in [\![t]\!] \tag{5.116}$$

Proof. The proof is conducted by structural induction over all terms; compare with Eqn. (3.1) *ff.* In the proof, all statements on known typings are justified by one of the typing rules (3.7) through (3.13) in Sect. 3.1.2 . For the sake of better readability, we do not want to explicitly name the respective typing rule in each case. Please use Sect. 3.1.2 as a permanent reference throughout the proof. Similarly, all statements on semantical equality of terms refer to and are justified by one of the semantic equations Eqns. (5.23) through (5.34). Again, we do not want to explicitly state which of the concrete semantic equations actually applies in each case.

For variables and constants, the proposition trivially holds. Given a variable $x : t$ we have that $\rho(x) \in [\![t]\!]$, because the environment ρ is, due to the definition of environments $\mathcal{E}$ in Def. 5.4, a type-respecting function. Similarly, for all constants $n_i : num$ we have that $[\![n_i]\!]_\rho$ is a function in $\mathbb{N}_0 \to [0,1]$

and therefore also $[\![n_i]\!]_\rho \in \lfloor \mathbb{N}_0 \to [0,1]_\omega \rfloor$ so that $[\![n_i]\!]_\rho \in [\![num]\!]$. A similar argument holds for the constants $\dot{t} : bool$ and $\dot{f} : bool$ so that $[\![\dot{t}]\!]_\rho \in [\![bool]\!]$ and $[\![\dot{f}]\!]_\rho \in [\![bool]\!]$.

Let us turn to the case of applications $(MN) : t_2$ with $M : t_1 \to t_2$ and $N : t_1$. The case is rather trivial. We can assume, as induction hypothesis, that $[\![M]\!]_\rho \in [[\![t_1]\!] \to [\![t_2]\!]]$ and $[\![N]\!]_\rho \in [\![t_1]\!]$. Now we know that that $[\![MN]\!]_\rho$ equals $[\![M]\!]_\rho\,[\![N]\!]_\rho$ and therefore $[\![MN]\!]_\rho \in [\![t_2]\!]$.

In case of recursions we deal with terms of the form $(\mu M) : t$ for some term $M : t \to t$. We know that $[\![\mu M]\!]_\rho$ equals $\Phi[\![M]\!]_\rho$. Due to the induction hypothesis we know that $[\![M]\!]_\rho \in [t \to t]$. Therefore, we know that $[\![M]\!]_\rho$ is an ω-continuous function in $[[\![t]\!] \to [\![t]\!]]$. Therefore, we can apply the fixed-point Theorem 2.79, which ensures that the fixed point of $[\![M]\!]_\rho$, i.e., $\Phi[\![M]\!]_\rho$, exists and furthermore $\Phi[\![M]\!]_\rho \in [\![t]\!]$.

Now, let us turn to probabilistic choices. Given a term $M|N$ with $M : t$ and $N : t$ for some higher type t. We know that $[\![M|N]\!]_\rho$ equals $[\![M]\!]_\rho \parallel^{0.5}_{0.5} [\![N]\!]_\rho$. Now, as induction hypothesis, we can assume that $[\![M]\!]_\rho \in [\![t]\!]$ and $[\![N]\!]_\rho \in [\![t]\!]$. Fortunately, we now have Corollary 5.19 to hand which ensures that $[\![M]\!]_\rho \parallel^{0.5}_{0.5} [\![N]\!]_\rho$ is an element of $[\![t]\!]$. A similar argumentation immediately applies to conditional expressions of the form *if* M *then* N_1 *else* N_2, because the semantics of conditional expressions is given completely in terms of the semantic probabilistic choice operator, compare with Eqn. (5.30).

Last but not least, let us turn to abstractions of the form $\lambda v.M$. We know that $(\lambda v.M) : (t_1 \to t_2)$ for a variable $v : t_1$ and a term $M : t_2$. Furthermore we know that $[\![\lambda v.M]\!]_\rho$ equals the function $d \in [\![t_1]\!] \mapsto [\![M]\!]_{\rho[v:=d]}$. Therefore, we know that $[\![\lambda v.M]\!]_\rho$ is an element of $[\![t_1]\!] \to [\![t_2]\!]$. However, we have not yet shown that $[\![\lambda v.M]\!]_\rho$ is also an element of $[[\![t_1]\!] \to [\![t_2]\!]]$, i.e., we do not yet know that $[\![\lambda v.M]\!] \in [\![t_1 \to t_2]\!]$. In order to have that, it remains to show that $[\![\lambda v.M]\!]_\rho$ is an ω-continuous function. This means that we need to show that for all ω-chains $c = (\, c_i \in [\![t_1]\!]\,)_{i\in\omega}$ we have that

$$[\![\lambda v.M]\!]_\rho(\sqcup_{[\![t_1]\!]}c) = \sqcup_{[\![t_2]\!]}([\![\lambda v.M]\!]^\dagger_\rho(c)) \tag{5.117}$$

Due to Eqn. (5.31) we know that $[\![\lambda v.M]\!]_\rho(\sqcup_{[\![t_1]\!]}c)$ equals

$$[\![M]\!]_{\rho[v:=\sqcup_{[\![t_1]\!]}c\,]} \tag{5.118}$$

Now, the l.u.b. substitution Lemma 5.21 applies and we know that Eqn. (5.118) equals

$$\sqcup_{[\![t_2]\!]}(([\![M]\!]_{\rho[v:=c_i]})_{i\in\omega}) \tag{5.119}$$

Now, again due to the definition of the semantic function in Eqn. (5.31) we know that Eqn. (5.119) equals

$$\sqcup_{[\![t_2]\!]}(([\![\lambda v.M]\!]_\rho(c_i))_{i\in\omega}) \tag{5.120}$$

Actually, we are already done now. Due to Def. 2.68 we know that Eqn. (5.120) equals

$$\sqcup_{[\![t_2]\!]}([\![\lambda v.M]\!]^{\dagger}_{\rho}(c)) \tag{5.121}$$

□

5.2.4 Futher Denotational Tools

In this section we provide two further, short technical lemmas that are needed later to prove the semantic correspondence for the probabilistic lambda calculus. The first, Lemma 5.23, allows, informally, for shortening environments. It states that the update to a variable enviroment does not affect the semantics of a term, as long as the updated variable is not free in the considered term. The second, Lemma 5.24, is a variable substitution lemma that is about the interplay of syntax, semantics and variable environments. If we are interested in the semantics of a term that results from a syntactical variable substitution in a core term, we can consider the semantics of only the core term instead, as long as we semantically update the variable environment appropriately. Given a term $M[v := N]$, where N is a program, its semantics with respect to an environment ρ equals the semantics of M with respect to the new environment $\rho[v := [\![N]\!]_{\emptyset}]$. Both lemmas are rather trivial and their proofs are straightforward by structural induction.

Lemma 5.23 (Shortening Environments) *For all terms $M : t$, environments $\rho \in \mathcal{E}$, data $d : [\![t]\!]$ and variables x that are not free in M, i.e., $x \notin V_{free}(M)$, we have that $[\![M]\!]_{\rho[x:=d]} = [\![M]\!]_{\rho}$*

Lemma 5.24 (Substitution Lemma) *For all terms M, environments $\rho : \mathcal{E}$, variables v and programs N we have that $[\![M[v := N]]\!]_{\rho} = [\![M]\!]_{\rho[v:=[\![N]\!]_{\emptyset}]}$*

5.3 Semantic Correspondence

In Theorem 5.55 we will prove the semantic correspondence between the denotational and the Markov chain semantics for ground types. Before we start, we introduce a useful notation $[c]$ for the retrieval of data points from the semantics of constants c that we also call data point semantics in the sequel.

Definition 5.25 (Data Point Semantics) *We define the data point semantics for all constants $n_i \in C_{num}$ and constants $\dot{t}, \dot{f} \in C_{bool}$, denoted by $[n_i]$, $[\dot{t}]$ and $[\dot{f}]$, as follows:*

$$[n_i] = i \tag{5.122}$$

$$[\dot{t}] = \mathcal{T} \tag{5.123}$$

$$[\dot{f}] = \mathcal{F} \tag{5.124}$$

Actually, given a constant c, we have that $[c]$ is just a shorthand notation for $[\![c]\!]_{\emptyset}^{-1}(1)$. Now, the semantic correspondence means that the probability of reducing a program M to a constant c is correctly yielded by applying the program denotation $[\![M]\!]_{\emptyset}$ to the data point $[c]$, i.e., the denotational semantics exactly mirrors the operational semantics for ground terms:

$$[\![M]\!]_{\emptyset}[c] = \big(M \Rightarrow c\big) \tag{5.125}$$

We investigate the semantic correspondence for ground types only, therefore we also speak about it as a basic correspondence. We will prove the semantic correspondence as a corollary of two restricted forms of correspondence, i.e., an upper bound correspondence $(M \Rightarrow c) \leqslant [\![M]\!]_{\emptyset}[c]$ which approximates the exact denotational semantics from above, and a lower bound correspondence $(M \Rightarrow c) \geqslant [\![M]\!]_{\emptyset}[c]$ which approximates the exact denotational semantics from below.

For deterministic calculi it is also usual to distinguish two kinds of relationships between operational and denotational semantics. For example, the text in [118] distinguishes between soundness and so-called computational adequacy. The soundness of the operational semantics with respect to the denotational semantics shows in the fact that $M \overset{\star}{\to} c$ implies $[\![M]\!]_{\emptyset} = [\![c]\!]$ for all programs M and constants c. Similarly, computational adequacy shows in the fact that $[\![M]\!]_{\emptyset} = [\![c]\!]$ implies $M \overset{\star}{\to} c$. In this terminology, the semantic correspondence is considered from the perspective of the denotational semantics, i.e., the denotational semantics is considered as given and soundness is a property of the operational semantics. This adheres to a viewpoint in which the denotational semantics comes first, i.e., as the specification of the programming language, and the operational semantics is considered as part of the implementation. In our terminology of upper and lower bound correspondence, the operational semantics is considered as given and it is the denotational semantics that provides a further characterization of the programming language semantics.

It is fair to say that upper bound correspondence is the probabilistic analogue of soundness, whereas lower-bound correspondence is the probabilistic analogue of computational adequacy. And, actually, for the deterministic fragment of the probabilistic lambda calculus the respective notions meet exactly. In the deterministic case, soundness is the easier to show, whereas computationally adequacy is the more sophisticated case. Similarly, upper bound correspondence is more straightforward to prove than lower bound correspondence.

Throughout the section we make use of the vector space laws available for the higher-type operators $\oplus$ and $\otimes$ as established by Lemma 5.10, however, without further mentioning them explicitly, because we are so familiar with them from the $\mathbb{R}$-vector space; compare with Sect. 2.5.5.

5.3.1 Upper Bound Correspondence

In order to prove upper bound correspondence, we first need a technical Lemma 5.26 that relates the one-step decomposition of the operational semantics as expressed by Eqn. (2.35) to the denotational semantics. In a sense, Lemma 5.26 expresses that the denotational semantics is preserved by the one-step semantics.

Lemma 5.26 (One-Step Semantic Correspondence) *For all closed lambda terms $P \in \Lambda_\emptyset$ we have that*

$$[\![P]\!]_\emptyset = \sum_{Q \in \Lambda_\emptyset} (P \rightarrow Q \otimes [\![Q]\!]_\emptyset) \tag{5.126}$$

Proof. The proof can be conducted by structural induction over terms M. We show the cases of constants, basic operators, conditional expressions, applications and deterministic choices only.

***In Case of Constants*:** As a first step, we can always transform the right-hand side of Eqn. (5.126) into the following equivalent formula:

$$(P \rightarrow P \otimes [\![P]\!]_\emptyset) \oplus \sum_{Q \neq P} (P \rightarrow Q \otimes [\![Q]\!]_\emptyset) \tag{5.127}$$

Now, Lemma 3.3 applies – see Eqn. (3.65). In case of constants $P \in C$ we know that $P \rightarrow P = 1$. It is clause (xviii), which applies due to its selector condition here, that allows us to conclude this. Furthermore, we know $P \rightarrow Q = 0$ for all $P \neq Q$. This time, it is clause (xx), i.e., the *else*-clause in Eqn. (3.65), that applies. Altogether, we therefore know that Eqn. (5.127) equals

$$(1 \otimes [\![P]\!]_\emptyset) \oplus \sum_{P \neq Q} (0 \otimes [\![Q]\!]_\emptyset) \tag{5.128}$$

And finally we know, due to Lemma 5.10, that Eqn. (5.128) equals $[\![P]\!]_\emptyset$.

Henceforth, in this proof, we will often conduct equivalent transformations of formulae that are valid due to the equalities that hold for the one-step semantics as expressed by Lemma 3.3 and Eqn. (3.65) without further mentioning this Lemma and; similarly, we often use the laws that hold for higher-type arithmetic as expressed by Lemma 5.10 without further mentioning Lemma 5.10. Lemma 5.10 establishes the vector space laws for the higher-type operators. Therefore, you might want to have Sect. 2.5.5 to hand as a reference in the sequel, which lists all the genuine and inherited vector space laws needed in this proof.

***In Case of Basic Operators*:** We prove Eqn. (5.126) for addition operator terms $P = +1(M)$ only. We need to distinguish two cases, i.e., the case that

M in $+1(M)$ is not a constant and the case that M is one of the constants n_i. We start with the case that M is not a constant. We need to prove the following corresponding instance of Eqn. (5.126):

$$[\![+1(M)]\!]_\emptyset = \sum_{Q \in \Lambda_\emptyset} (+1(M) \rightarrow Q \otimes [\![Q]\!]_\emptyset) \tag{5.129}$$

The right-hand side of Eqn. (5.129) equals the following:

$$\sum_{M' \in \Lambda_\emptyset} (+1(M) \rightarrow +1(M') \otimes [\![+1(M')]\!]_\emptyset) \oplus \sum_{Q \notin \{+1(M') \,|\, M' \in \Lambda_\emptyset\}} (+1(M) \rightarrow Q \otimes [\![Q]\!]_\emptyset) \tag{5.130}$$

With a similar argument as in the case of constants we can see that the second summand in Eqn. (5.130) vanishes. Again, Eqn. (3.65) applies, where the side conditions $M \notin C$ are crucial to uniquely identify the correct clause (xx). Similarly, due to clause (i) we know that $+1(M) \rightarrow +1(M')$ equals $M \rightarrow M'$. Altogether this means that Eqn. (5.130) equals

$$\sum_{M' \in \Lambda_\emptyset} (M \rightarrow M' \otimes [\![+1(M')]\!]_\emptyset) \tag{5.131}$$

Now, due to the definition of the basic operator $[\![+1]\!]$ in Def. 5.6 plus the distributivity of function application over the higher-type arithmetic operators as expressed in Lemma 5.13, we have, after a series of transformations, that 5.131 equals

$$n \in \mathbb{N}_0 \mapsto \begin{cases} \left(\sum_{M' \in \Lambda_\emptyset} (M \rightarrow M' \otimes [\![M']\!]_\emptyset)\right)(n-1) & , n > 0 \\ 0 & , else \end{cases} \tag{5.132}$$

Now, the induction hypothesis applies so that Eqn. (5.132) equals

$$n \in \mathbb{N}_0 \mapsto \begin{cases} [\![M]\!](n-1) & , n > 0 \\ 0 & , else \end{cases} \tag{5.133}$$

Finally, we know that Eqn. (5.133) equals $[\![+1(M)]\!]$.

Next, we turn to the case that M in $+1(M)$ is a constant n_i for some $i \in \mathbb{N}_0$. In such a case, we need to prove the following:

$$[\![+1(n_i)]\!]_\emptyset = \sum_{Q \in \Lambda_\emptyset} (+1(n_i) \rightarrow Q \otimes [\![Q]\!]_\emptyset) \tag{5.134}$$

The right-hand side of Eqn. (5.134) equals the following:

$$(+1(n_i) \rightarrow n_{i+1} \otimes [\![n_{i+1}]\!]_\emptyset) \oplus \sum_{Q \in \Lambda_\emptyset \backslash \{n_{i+1}\}} (+1(n_i) \rightarrow Q \otimes [\![Q]\!]_\emptyset) \tag{5.135}$$

Again, we can exploit Eqn. (3.65), clauses (ii) and (xx) this time, and we see that Eqn. (5.135) equals $[\![n_{i+1}]\!]$. Therefore, it suffices to show that $[\![n_{i+1}]\!]$ equals $[\![+1(n_i)]\!]$ in order to complete the proof. Now, by Def. 5.6 we have that

$$[\![+1(n_i)]\!]_\rho = n \in \mathbb{N}_0 \mapsto \begin{cases} [\![n_i]\!]_\rho(n-1) & , n > 0 \\ 0 & , else \end{cases} \tag{5.136}$$

Now, by the definition of the semantics for number constants $[\![n_i]\!]$ in Eqn. (5.24) we know that Eqn. (5.136) equals

$$n \in \mathbb{N}_0 \mapsto \begin{cases} k \in \mathbb{N}_0 \mapsto \left\{ \begin{matrix} 1 & , k = i \\ 0 & , else \end{matrix} \right\} (n-1) & , n > 0 \\ 0 & , else \end{cases} \tag{5.137}$$

Now, by function application and a further simple transformation of $n - 1 = i$ into $n = i + 1$ we have that Eqn. (5.137) equals

$$n \in \mathbb{N}_0 \mapsto \begin{cases} \left\{ \begin{cases} 1 & , n = i + 1 \\ 0 & , else \end{cases} \right\} & , n > 0 \\ 0 & , else \end{cases} \tag{5.138}$$

Because we have that $i+1 > 0$ for all i we can rewrite Eqn. (5.138) as follows:

$$n \in \mathbb{N}_0 \mapsto \begin{cases} 1 & , n = i + 1 \\ 0 & , else \end{cases} \tag{5.139}$$

Finally, again due to Eqn. (5.24) we have that Eqn. (5.139) equals $[\![n_{(i+1)}]\!]$.

***In Case of Conditional Expressions*:** In case of conditional expressions of the form $P = \mathit{if}\, M \; \mathit{then}\; N' \; \mathit{else}\; N''$ we again distinguish two cases, i.e., the case that the condition M is one of the constants $\dot{t}$ or $\dot{f}$ or otherwise that M is not a constant. We show only the case that M is not a constant. After exploiting Eqn. (3.65), clause (ix), we know that it suffices to show the following:

$$[\![\mathit{if}\, M \; \mathit{then}\; N' \; \mathit{else}\; N'']\!]_\emptyset = \sum_{M' \in \Lambda_\emptyset} (M \to M' \otimes [\![\mathit{if}\, M' \; \mathit{then}\; N' \; \mathit{else}\; N'']\!]_\emptyset) \tag{5.140}$$

Now, due to the semantic equation Eqn. (5.30) we know that the right-hand side of Eqn. (5.140) equals

$$\sum_{M' \in \Lambda_\emptyset} (M \to M' \otimes ([\![M']\!]_\rho(\mathcal{T}) \otimes [\![N']\!]_\rho \;\oplus\; [\![M']\!]_\rho(\mathcal{F}) \otimes [\![N'']\!]_\rho)) \tag{5.141}$$

After several transformations we have that Eqn. (5.141) equals

$$\Big(\sum_{M' \in \Lambda_\emptyset}(M \to M' \otimes [\![M']\!]_\rho)\Big)(\mathcal{T}) \otimes [\![N']\!]_\rho \;\oplus\; \Big(\sum_{M' \in \Lambda_\emptyset}(M \to M' \otimes [\![M']\!]_\rho)\Big)(\mathcal{F}) \otimes [\![N'']\!]_\rho \tag{5.142}$$

Now, the induction hypothesis applies so that Eqn. (5.142) equals

$$[\![M]\!]_\rho(\mathcal{T}) \otimes [\![N']\!]_\rho \;\oplus\; [\![M]\!]_\rho(\mathcal{F}) \otimes [\![N'']\!]_\rho \tag{5.143}$$

Finally, we know that Eqn. (5.143) equals $[\![\mathit{if}\, M \;\mathit{then}\; N' \;\mathit{else}\; N'']\!]$.

In Case of Applications: In case of applications $P = MN$ we need to distinguish two cases, i.e., the case in which M equals an abstraction $\lambda v.B$ and the case in which M does not equal an abstraction. We show only the case that M does not equal an abstraction. After exploiting Eqn. (3.65), clause (xii), we know that it suffices to show the following:

$$[\![MN]\!]_\emptyset = \sum_{M' \in \Lambda_\emptyset} (M \to M' \otimes [\![M'N]\!]_\emptyset) \tag{5.144}$$

Now, due to the semantic equation Eqn. (5.32) we know that the right-hand side of Eqn. (5.144) equals

$$\sum_{M' \in \Lambda_\emptyset} (M \to M' \otimes ([\![M']\!]_\emptyset [\![N]\!]_\emptyset)) \tag{5.145}$$

Next, we can exploit the distributiveness of higher-type arithmetic operators over and function application in Lemma 5.13. Therefore, first due to Eqn. (5.49) and second due to Eqn. (5.50), we have that Eqn. (5.145) equals

$$\Big(\sum_{M' \in \Lambda_\emptyset} (M \to M' \otimes [\![M']\!]_\emptyset)\Big) [\![N]\!]_\emptyset \tag{5.146}$$

Now, the induction hypothesis applies so that Eqn. (5.146) equals $[\![M]\!]_\emptyset [\![N]\!]_\emptyset$ which in turn equals $[\![MN]\!]_\emptyset$, again due to Eqn. (5.32).

In Case of Deterministic Choice: In case of deterministic choices of the form $P = M|N$ we need to distinguish the cases in which M is different from N from the cases in which M equals N. We show only the case that $M \neq N$. We need to show the following:

$$[\![M|N]\!]_\emptyset = \sum_{Q \in \Lambda_\emptyset} (M|N \to Q \otimes [\![Q]\!]_\emptyset) \tag{5.147}$$

Now, in any case the right-hand side of Eqn. (5.147) can be transformed into the following equivalent formula:

$$(M|N \rightarrow M \otimes [\![M]\!]_\emptyset) \oplus (M|N \rightarrow N \otimes [\![N]\!]_\emptyset) \oplus \sum_{Q \in \Lambda_\emptyset \setminus \{M,N\}} (M|N \rightarrow Q \otimes [\![Q]\!]_\emptyset) \quad (5.148)$$

Due to Eqn. (3.65) we know that Eqn. (5.148) equals

$$(0.5 \otimes [\![M]\!]_\emptyset) \ \oplus \ (0.5 \otimes [\![N]\!]_\emptyset) \quad (5.149)$$

Finally, due to the semantic equation Eqn. (5.34) we have that Eqn. (5.149) equals $[\![M|N]\!]_\emptyset$. □

Next, we prove the upper bound correspondence in Lemma 5.27. Upper bound correspondence means that the denotational semantics approximates the Markov chain semantics from above, i.e., $(M \Rightarrow c) \leqslant [\![M]\!]_\emptyset[c]$. Upper bound correspondence means that the investigated denotational semantics yields an upper bound for the intended, exact denotational semantics, which we expect, pointwise, to equal the evaluation semantics, i.e., $(M \Rightarrow c) = [\![M]\!]_\emptyset[c]$.

Lemma 5.27 (Upper Bound Correspondence) *The denotational semantics approximates the evaluation semantics from above. For each program M and constant c we have that*

$$(M \Rightarrow c) \leqslant [\![M]\!]_\emptyset[c] \quad (5.150)$$

Proof. We will show Eqn. (5.150) as a corollary of a more general result which shows that the denotational semantics is preserved by the reduction semantics for all closed terms, i.e., not only for programs; see Eqn. (5.153). As a first step, we can see that $(i \otimes [\![c]\!]_\emptyset)[c]$ equals i for each $i \in \mathbb{R}$; compare with Def. 5.8 and the semantic equations Eqns. (5.24), (5.25) and (5.26). Therefore, we have that Eqn. (5.150) is equivalent to

$$((M \Rightarrow c) \otimes [\![c]\!]_\emptyset)[c] \ \leqslant \ [\![M]\!]_\emptyset[c] \quad (5.151)$$

Now, due to the pointwise definition of basic semantic objects we also have that Eqn. (5.151) equals

$$(M \Rightarrow c) \otimes [\![c]\!]_\emptyset \ \sqsubseteq \ [\![M]\!]_\emptyset \quad (5.152)$$

Now, we have that Eqn. (5.152) is a special case of the following, more general proposition for all closed terms N:

$$(M \Rightarrow N) \otimes [\![N]\!]_\emptyset \ \sqsubseteq \ [\![M]\!]_\emptyset \quad (5.153)$$

We proceed by proving that the more general proposition Eqn. (5.153) holds true for all closed terms M and N. We start with proving that all finite approximations to Eqn. (5.153) hold true, i.e., that for all $n \in \mathbb{N}_0$ we have the following:

$$\eta_S^n\langle M, N\rangle \otimes [\![N]\!]_\emptyset \ \sqsubseteq \ [\![M]\!]_\emptyset \quad (5.154)$$

We prove Eqn. (5.154) by natural induction over n.

In Case of $\boldsymbol{n = 0}$**:** We need to the distinguish the cases that $M = N$ and $M \neq N$. In case $M = N$ we know by the decomposition of bounded hitting probabilities, i.e., Lemma 2.31, Eqn. (2.37), that $\eta_S^0\langle M, N\rangle = 1$ so that we have that $1 \otimes [\![N]\!]_\emptyset$ equals $[\![M]\!]_\emptyset$ and therefore also approximates $[\![M]\!]_\emptyset$. Similarly, in case $M \neq N$ we know by Lemma 2.31, Eqn. (2.38), that $\eta_S^0\langle M, N\rangle = 0$. Therefore, given that the type of M and N is T, we know that $0 \otimes [\![N]\!]_\emptyset$ equals $\bot_T$ and therefore approximates $[\![M]\!]_\emptyset$.

In Case of $\boldsymbol{n \geqslant 1}$**:** We start with the left-hand side of Eqn. (5.154), i.e.,

$$\eta_S^n\langle M, N\rangle \otimes [\![N]\!]_\emptyset \tag{5.155}$$

Due to Lemma 2.31, Eqn. (2.39), we know that Eqn. (5.155) equals

$$\Big(\sum_{M' \in \Lambda_\emptyset} M \to M' \cdot \eta_S^{n-1}\langle M', N\rangle\Big) \otimes [\![N]\!]_\emptyset \tag{5.156}$$

Furthermore, we know that Eqn. (5.156) equals

$$\sum_{M' \in \Lambda_\emptyset} (M \to M' \otimes (\eta_S^{n-1}\langle M', N\rangle \otimes [\![N]\!]_\emptyset)) \tag{5.157}$$

Now, we know that $\eta_S^{n-1}\langle M', N\rangle \otimes [\![N]\!]_\emptyset \sqsubseteq [\![M']\!]_\emptyset$, because the induction hypothesis applies. Due to the ω-continuity of functional scalar multiplication, Lemma 5.14, we know that $\otimes$ is monotone in its second argument, compare with Def. 2.70, and therefore we have that Eqn. (5.157) is less than or equal to

$$\sum_{M' \in \Lambda_\emptyset} M \to M' \otimes [\![M']\!]_\emptyset \tag{5.158}$$

Finally, due to Lemma 5.26 we know that Eqn. (5.158) equals $[\![M]\!]_\emptyset$, which proves Eqn. (5.154).

Now, due to Eqn. (5.154) we know that $[\![M]\!]_\emptyset$ is an upper bound of the chain $(\eta_S^n\langle M, N\rangle \otimes [\![N]\!]_\emptyset)_{n \in \omega}$. Furthermore, by the definition of reduction probabilities in Def. 3.5, the limit of bounded hitting probabilities as characterized in Corollary 2.27 and the ω-continuity of $\otimes$ as stated by Lemma 5.14 we know that $(M \Rightarrow N) \otimes [\![N]\!]_\emptyset$ is the least upper bound of the chain $(\eta_S^n\langle M, N\rangle \otimes [\![N]\!]_\emptyset)_{n \in \omega}$. Therefore, we know that Eqn. (5.153) holds true. □

5.3.2 Lower Bound Correspondence

In order to prove lower bound correspondence we first need to introduce a stronger property for lambda terms that faithfully generalizes lower bound correspondence from programs to open terms as well as terms of higher type. We will then proceed to prove this stronger property for all lambda terms by structural induction. With this proof idea we follow the original approach of proving the denotational semantics of PCF (programming language for computable functions) in [223] complete for programs. We call the property α that we need to establish *over-performance* or performance for short.

Definition 5.28 (Performing Term) *We define the property $\alpha(M)$ for all terms $M \in \Lambda$ by the following comprehensive cases:*

(a) If M is a program then $\alpha(M)$ **iff** *$(M \Rightarrow c) \geqslant [\![M]\!]_\emptyset([c])$ for all $c \in C$.*
(b) If M is a closed lambda term of higher type $t_1 \rightarrow t_2$ then $\alpha(M)$ **iff** *$\alpha(MN)$ for each closed lambda term N of type t_1 for which $\alpha(N)$ holds.*
(c) If M is an open term with $FV(M) = \{x_1, \ldots, x_n\}$ then $\alpha(M)$ **iff** *$\alpha(M[x_1 := N_1] \ldots [x_n := N_n])$ for each collection of closed lambda terms $N_1, \ldots, N_n$ for which $\alpha(N_1)$ through $\alpha(N_n)$ hold.*

In the sequel, we will prove that all terms M are performing, i.e., $\alpha(M)$, by a structural induction. Due to its length we split this proof into a sequence of technical lemmas in the sequel. There will be one lemma for each case of term construction, proving the performance of a closed term based on the assumption that its component terms, i.e., the terms that it is constructed from are already performing. Then, the collection of these lemmas can be used to constitute a correct structural inductive proof of the desired property. These lemmas are accompanied by further technical lemmas dealing with big-step semantics or big-step semantical approximations concerning the respective case. The case of recursion is particularly work-intensive to prove. Here, we need the further notions of syntactical approximation and respective lemmas. We start with the case of values, which encompass constants and lambda abstractions.

Lemma 5.29 (Bounded Big-Step Semantics of Value Expressions) *Given values $V \in \Lambda_V$, $V' \in \Lambda_V$ and a number $n \in \mathbb{N}_0$ we have that*

$$\eta^n_S\langle V, V'\rangle = \begin{cases} 1 & , V = V' \\ 0 & , else \end{cases} \tag{5.159}$$

Proof. We distinguish the cases in which $V = V'$ from cases in which $V \neq V'$. Case $V = V'$ immediately holds, see Lemma 2.31, Eqn. (2.37). Case $V \neq V'$ is shown by natural induction over the bound n.

***In Case of n=0*:** In case $n = 0$ we know that $\eta^0_S\langle V, V'\rangle = 0$ by the definition of bounded hitting probabilities Def. 2.24; see also the definition of first hitting times in Def. 2.22.

***In Case of $n \geqslant 1$*:** In case $n \geqslant 1$ we have, due to the one-step decomposition of bounded hitting probabilities, i.e., Lemma 2.31, Eqn. (2.39), that

$\eta_S^{n+1}\langle V, V'\rangle$ equals

$$\sum_{M \in \Lambda} V \to M \cdot \eta_S^{n+1}\langle M, V'\rangle \tag{5.160}$$

Now, Eqn. (5.160) can always be transformed into

$$V \to V \cdot \eta_S^{n}\langle V, V'\rangle \ + \sum_{M \in \Lambda\backslash\{V\}} V \to M \cdot \eta_S^{n+1}\langle M, V'\rangle \tag{5.161}$$

Due to the induction hypothesis we know that the first summand in Eqn. (5.161) equals zero. Due to Lemma 3.3, Eqn. (3.65), clause (xx), we know that $V \to M$ equals zero for all values $V \in \Lambda_V$ and terms $M \notin \Lambda\backslash\{V\}$ so that the second summand in Eqn. (5.161) also equals zero, which completes the proof. □

Lemma 5.30 (Big-Step Semantics of Value Expressions) *Given values $V \in \Lambda_V$ and $V' \in \Lambda_V$ we have that*

$$V \Rightarrow V' = \begin{cases} 1 & , V = V' \\ 0 & , else \end{cases} \tag{5.162}$$

Proof. This Lemma follows as an immediate corollary from Lemma 5.29. Due to the definition of reduction probabilities in Def. 3.5 we have that $V \Rightarrow V'$ equals $\lim\limits_{n\to\infty} \eta_S^n\langle V, V'\rangle$. Now, by Lemma 5.29 we have that $\eta_S^n\langle V, V'\rangle$ yields the correct right-hand side value of Eqn. (5.162) as a constant value, which proves Eqn. (5.162). □

Lemmas 5.29 and 5.30 hold, because a reduction stabilizes after it has reached a value. In the sequel, as a helper Lemma, we also need to characterize how performance transports to expressions of higher type in general, as expressed by Lemma 5.31.

Lemma 5.31 (Performance of Higher-Type Expressions) *Given a closed term of higher type $M : t_1 \to t_2$, closed terms $\widetilde{N} = N_0, \ldots, N_n$ such that $\alpha(N_i)$ for all $0 \leqslant i \leqslant n$ and a program $M\widetilde{N}$. Then, we have that $\alpha(M\widetilde{N})$ implies $\alpha(M)$.*

Proof. This Lemma immediately follows by the definition of performing terms Def. 5.28, by applying clause (b) of Def. 5.28 n times plus once to $MN_0 \ldots N_n$, i.e., for each $1 \leqslant i \leqslant n$ we can conclude by clause (b) that $\alpha(MN_0 \ldots N_i)$ implies $\alpha(MN_0 \ldots N_{i-1})$ and, finally, that $\alpha(MN_0)$ implies $\alpha(M)$. □

Performance of Basic Operators

Next, we turn to the case of basic operators. Here we first introduce a big-step semantics in Lemma 5.32. Then, on the basis of this big-step semantics, we establish the performance of basic operators in Lemma 5.33.

Lemma 5.32 (Big-Step Semantics of Basic Operators) *Given a closed term M:num of the ground type of numbers we have that*

$$+1(M) \Rightarrow n_i = \begin{cases} M \Rightarrow n_{i-1} & , i > 0 \\ 0 & , else \end{cases} \tag{5.163}$$

$$-1(M) \Rightarrow n_i = \begin{cases} M \Rightarrow n_{i+1} & , i > 0 \\ (M \Rightarrow n_0) + (M \Rightarrow n_1) & , else \end{cases} \tag{5.164}$$

$$0?(M) \Rightarrow b = \begin{cases} M \Rightarrow n_0 & , b = \dot{t} \\ \eta_S\langle M, C\backslash\{n_0\}\rangle & , b = \dot{f} \end{cases} \tag{5.165}$$

Proof. We show this only for lambda terms $+1(M)$. The other cases can be proven analogously. First, due to the definition of reduction probabilities in Def. 3.5 and Corollary 2.27 we have that Eqn. (5.163) is equivalent to

$$\lim_{n\to\infty} \eta_S^n\langle +1(M), n_i\rangle = \begin{cases} \lim_{n\to\infty} \eta_S^n\langle M, n_{i-1}\rangle & , i > 0 \\ 0 & , else \end{cases} \tag{5.166}$$

Due to the monotonicity of bounded hitting probabilities, i.e., Lemma 2.28; and the monotone convergence Theorem 2.93 we have that $\lim_{n\to\infty} \eta_S^n\langle M, n_{i-1}\rangle$, i.e., the right hand side of Eqn. (5.166), is also the least upper bound of its sequence $(\eta_S^n\langle M, n_{i-1}\rangle)_{n\in\omega}$. Similarly we know that the left-hand side of Eqn. (5.166), i.e., $\lim_{n\to\infty} \eta_S^n\langle +1(M), n_i\rangle$, equals the least upper bound of $(\eta_S^n\langle +1(M), n_i\rangle)_{n\in\omega}$. For the same reasons, we know that $\lim_{n\to\infty} \eta_S^n\langle +1(M), n_i\rangle$ equals $\lim_{n\to\infty} \eta_S^{n+m}\langle +1(M), n_i\rangle$ for an arbitrary but fixed $m \in \mathbb{N}_0$. In particular, we know that $\lim_{n\to\infty} \eta_S^n\langle +1(M), n_i\rangle$ equals $\lim_{n\to\infty} \eta_S^{n+1}\langle +1(M), n_i\rangle$. Altogether, we therefore have that, in order to prove Eqn. (5.166) it suffices to prove that the following holds for all $n \in \mathbb{N}_0$:

$$\eta_S^{n+1}\langle +1(M), n_i\rangle = \begin{cases} \eta_S^n\langle M, n_{i-1}\rangle & , i > 0 \\ 0 & , else \end{cases} \tag{5.167}$$

We will prove Eqn. (5.167) by natural induction over the bound $n \in \mathbb{N}_0$. In the proof we will maintain the distinction of the case that M equals a number constant $n_j \in C$ from the case that M is not a constant, i.e., $M \notin C$. Actually, in case of constants, Eqn. (5.167) can be proven immediately, without natural induction, and we will do so as preparatory work now. First, due to the one-step decomposition of bounded hitting probabilities, i.e., Lemma 2.31, we have for all constants $n_j \in C_{num}$ and bounds $n \in \mathbb{N}_0$ that $\eta_S^{n+1}\langle +1(n_j), n_i\rangle$ equals the following:

$$\sum_{Q\in\Lambda_\emptyset} +1(n_j) \to Q \cdot \eta_S^n\langle Q, n_i\rangle \tag{5.168}$$

We can transform Eqn. (5.168) into the following:

$$+1(n_j) \to n_{j+1} \cdot \eta_S^n\langle n_i, n_i\rangle + \sum_{Q \in \Lambda_\emptyset \setminus \{n_i\}} +1(n_j) \to Q \cdot \eta_S^n\langle Q, n_i\rangle \tag{5.169}$$

Due to Lemma 3.3, Eqn. (3.65), clause (xx), we know that $+1(n_j) \to Q$ in the second summand of Eqn. (5.169) equals zero for all $Q \neq n_{j+1}$. Due to the definition of bounded hitting probabilities in Def. 2.24 we know that $\eta_S^n\langle M, M\rangle$ equals one for all terms $M \in \Lambda_\emptyset$ and therefore also $\eta_S^n\langle n_i, n_i\rangle$ in the first summand of Eqn. (5.169) equals one. Altogether we therefore have that

$$\eta_S^{n+1}\langle +1(n_j), n_i\rangle = \begin{cases} 1 & , n_{j+1} = n_i \\ 0 & , n_{j+1} \neq n_i \end{cases} \tag{5.170}$$

Due to the definition of number constants in Sect. 3.1.1 and a further case distinction, it is possible to rewrite Eqn. (5.170) into the following form:

$$\eta_S^{n+1}\langle +1(n_j), n_i\rangle = \begin{cases} \begin{cases} 1 & , n_j = n_{i-1} \\ 0 & , n_j \neq n_{i-1} \end{cases} & , i > 0 \\ 0 & , else \end{cases} \tag{5.171}$$

Remember that constants are values in the sense of Def. 3.1.3. Therefore Lemma 5.29 applies and we have that Eqn. (5.171) equals

$$\eta_S^{n+1}\langle +1(n_j), n_i\rangle = \begin{cases} \eta_S^n\langle n_j, n_{i-1}\rangle & , i > 0 \\ 0 & , else \end{cases} \tag{5.172}$$

Note that Eqn. (5.172) just proves Eqn. (5.167) for the case of constants. Let us do some further preparatory work and turn our focus onto $\eta_S^{n+1}\langle +1(M), n_i\rangle$ for terms $M \notin C_{num}$. For all terms $M \notin C_{num}$ and bounds $n \in \mathbb{N}_0$ we have that $\eta_S^{n+1}\langle +1(M), n_i\rangle$ equals

$$\sum_{Q \in \Lambda_\emptyset} +1(M) \to Q \cdot \eta_S^n\langle Q, n_i\rangle \tag{5.173}$$

We can transform Eqn. (5.173) into the following:

$$\sum_{M' \in \Lambda_\emptyset} +1(M) \to +1(M') \cdot \eta_S^n\langle +1(M'), n_i\rangle + \sum_{Q \notin \{+1(M') \,|\, M' \in \Lambda_\emptyset\}} +1(M) \to Q \cdot \eta_S^n\langle Q, n_i\rangle \tag{5.174}$$

In case M is not a constant we know, due to Lemma 3.3, Eqn. (3.65), clause (i), that $+1(M) \to +1(M')$ equals $M \to M'$. Furthermore, due to Lemma 3.3, Eqn. (3.65), clause (xx), we know that $+1(M) \to Q$ equals zero for all terms $Q \in \Lambda_\emptyset$ that are different from $+1(M')$ for all terms $M' \in \Lambda_\emptyset$. Therefore, we have that Eqn. (5.174) implies the following for all terms $M \notin C_{num}$ and bounds $n \in \mathbb{N}_0$:

$$\eta_S^{n+1}\langle +1(M), n_i\rangle = \sum_{M' \in \Lambda_\emptyset} M \to M' \cdot \eta_S^n\langle +1(M'), n_i\rangle \tag{5.175}$$

We proceed with the natural induction for Eqn. (5.167) over bound $n \in \mathbb{N}_0$.

***In Case of n=0*:** The case that M is a constant is already proven by Eqn. (5.172). Let us turn to the case of terms $M \notin C_{num}$ that are not number constants. This case is also almost trivial. Informally it is not possible to reach a constant in just one step in this case. Informally we have, due to Eqn. (5.175), that

$$\eta_S^1\langle +1(M), n_i\rangle = \sum_{M' \in \Lambda_\emptyset} M \to M' \cdot \eta_S^0\langle +1(M'), n_i\rangle \tag{5.176}$$

Now, $+1(M')$ is not a constant and therefore, due to the definition of bounded hitting probabilities, Def. 2.24, we know that $\eta_S^0\langle +1(M'), n_i\rangle$ equals zero. Therefore, Eqn. (5.176) implies that $\eta_S^1\langle +1(M), n_i\rangle$ equals zero. On the other hand, we have that $\eta_S^0\langle M, n_i\rangle$ also equals zero, because M is not a constant and therefore Lemma 3.3, Eqn. (3.65), clause (xx) applies, which completes the proof of Eqn. (5.167) in this case.

***In Case of n ⩾ 1*:** Again, the case that M is a constant is already proven by Eqn. (5.172). In case $M \in C_{num}$ is not a constant, we have, due to Eqn. (5.175), that

$$\eta_S^{n+2}\langle +1(M), n_i\rangle = \sum_{M' \in \Lambda_\emptyset} M \to M' \cdot \eta_S^{n+1}\langle +1(M'), n_i\rangle \tag{5.177}$$

Now, the induction hypothesis applies to $\eta_S^{n+1}\langle +1(M'), n_i\rangle$ so that Eqn. (5.177) is equivalent to

$$\eta_S^{n+2}\langle +1(M), n_i\rangle = \sum_{M' \in \Lambda_\emptyset} M \to M' \cdot \begin{cases} \eta_S^n\langle M', n_i\rangle & , i > 0 \\ 0 & , else \end{cases} \tag{5.178}$$

Now, we have that n_i is arbitrary but fixed in Eqn. (5.178) so that Eqn. (5.178) can be rewritten as follows:

$$\eta_S^{n+2}\langle +1(M), n_i\rangle = \begin{cases} \sum_{M' \in \Lambda_\emptyset} M \to M' \cdot \eta_S^n\langle M', n_i\rangle & , i > 0 \\ 0 & , else \end{cases} \tag{5.179}$$

As the final step, the one-step decomposition of bounded hitting probabilities, Lemma 2.31, can be applied backwards to the branch $i > 0$ in Eqn. (5.179) so that Eqn. (5.179) is equivalent to

$$\eta_S^{n+2}\langle +1(M), n_i\rangle = \begin{cases} \eta_S^{n+1}\langle M, n_i\rangle & , i > 0 \\ 0 & , else \end{cases} \tag{5.180}$$

□

Lemma 5.33 (Performance of Basic Operators) *Given a performing, closed term of the ground type of numbers, i.e., $M{:}num$ such that $\alpha(M)$, we have that $\alpha(+1(M))$, $\alpha(-1(M))$ and $\alpha(0?(M))$.*

Proof. We show this only for lambda terms $+1(M)$. The other cases are analogous. We start with the right-hand side of Def. 5.28, clause (a), for an arbitrary constant $c \in C_{num}$ as follows:

$$[\![+1(M)]\!]_{\emptyset}([n_i]) = \begin{cases} [\![M]\!]([n_i]-1) & ,[n_i] > 0 \\ 0 & ,else \end{cases} \tag{5.181}$$

Due to the definition of data points we have that $[n_i] = i$ and, again due to the definition of data points, twice, we have that $[n_i]-1$ equals $[n_{i-1}]$. Therefore, and due to the premise $\alpha(M)$, we know that Eqn. (5.181) implies

$$[\![+1(M)]\!]_{\emptyset}([n_i]) \leqslant \begin{cases} M \Rightarrow n_{i-1} & ,i > 0 \\ 0 & ,else \end{cases} \tag{5.182}$$

Now, due to Lemma 5.32 we have that Eqn. (5.182) implies

$$[\![+1(M)]\!]_{\emptyset}([n_i]) \leqslant M \Rightarrow n_i \tag{5.183}$$

Eqn. (5.183) completes the proof, because it fulfills the definition of performing terms Def. 5.28, clause (a). □

Performance of Conditional Expressions

We proceed with closed conditional expressions. We introduce a big-step semantics for conditional expressions in Lemma 5.34 and use it to prove the performance of conditional expressions in Lemma 5.35.

Lemma 5.34 (Big-Step Approximation of Conditional Expressions) *Given a closed term $if(B,N',N'')$, closed terms $\widetilde{N} = N_0,\ldots,N_n$ such that $if(B,N',N'')\widetilde{N}$ is a program and a constant $c \in C$ we have that*

$$if(B,N',N'')\widetilde{N}\Rightarrow c \;\geqslant\; B\Rightarrow\dot{t}\cdot N'\widetilde{N}\Rightarrow c \,+\, B\Rightarrow\dot{f}\cdot N''\widetilde{N}\Rightarrow c \tag{5.184}$$

Proof. By the definition of reduction probabilities in Def. 3.5 we have that the right-hand side of Eqn. (5.184) is equivalent to

$$\eta_S\langle B,\dot{t}\rangle\cdot\eta_S\langle N'\widetilde{N},c\rangle + \eta_S\langle B,\dot{f}\rangle\cdot\eta_S\langle N''\widetilde{N},c\rangle \tag{5.185}$$

Due to Corollary 2.27 and the properties of limits of sequences in Lemma 2.91 we have that Eqn. (5.185) is equivalent to

$$\lim_{n\to\infty}\left(\,\eta_S^n\langle B,\dot{t}\rangle\cdot\eta_S\langle N'\widetilde{N},c\rangle + \eta_S^n\langle B,\dot{f}\rangle\cdot\eta_S\langle N''\widetilde{N},c\rangle\,\right) \tag{5.186}$$

Due to the monotonicity of bounded hitting probabilities, i.e., Lemma 2.28, the monotonicity of real number multiplication and addition, and the monotone convergence Theorem 2.93 we have that the limit Eqn. (5.186) is also the least upper bound of its sequence. Therefore, in order to prove Eqn. (5.184) it suffices to prove that the following holds for all $n \in \mathbb{N}_0$:

$$\mathit{if}(B, N', N'')\widetilde{N} \Rightarrow c \geqslant \eta_S^n\langle B, \dot{t}\rangle \cdot \eta_S\langle N'\widetilde{N}, c\rangle + \eta_S^n\langle B, \dot{f}\rangle \cdot \eta_S\langle N''\widetilde{N}, c\rangle \quad (5.187)$$

We prove Eqn. (5.187) by natural induction over n, i.e., the bound common to all bounded hitting probabilities in Eqn. (5.187).

***In Case of n=0*:** We can distinguish the three cases that the condition B is one of the truth constants $\dot{t}$ or $\dot{f}$ or otherwise neither of the truth constants. The proofs for the cases $B = \dot{t}$ and $B = \dot{f}$ are almost the same and we show the case $B = \dot{t}$ only. In case $B = \dot{t}$ the left-hand side of Eqn. (5.187) takes the following form:

$$\mathit{if}(\dot{t}, N', N'')\widetilde{N} \Rightarrow c \quad (5.188)$$

Due to Eqn. (2.35) we know that Eqn. (5.188) equals the following:

$$\sum_{M' \in \Lambda_P} (\mathit{if}(\dot{t}, N', N'')\widetilde{N} \to M') \cdot \eta_S\langle M', c\rangle \quad (5.189)$$

Selecting one particular summand of Eqn. (5.189) we have that Eqn. (5.189) is greater than or equal ($\geqslant$) to

$$(\mathit{if}(\dot{t}, N', N'')\widetilde{N} \to N'\widetilde{N}) \cdot \eta_S\langle N'\widetilde{N}, c\rangle \quad (5.190)$$

Due to Eqn. (3.65), clause (x), we know that $(\mathit{if}(\dot{t}, N', N'') \to N')$ equals one. Therefore, due to Eqn. (3.65), clause (xii), n times, we know that also $(\mathit{if}(\dot{t}, N', N'')\widetilde{N} \to N'\widetilde{N})$ equals one. Therefore, we know that Eqn. (5.190) equals

$$\eta_S\langle N'\widetilde{N}, c\rangle \quad (5.191)$$

Now, we have that $\eta_S^0\langle \dot{t}, \dot{t}\rangle = 1$ and $\eta_S^0\langle \dot{t}, \dot{f}\rangle = 0$ due to Lemma 3.3, Eqn. (3.65), clause (xviii) resp. clause (xx). Therefore, we finally know that Eqn. (5.191) equals the right-hand side of Eqn. (5.187), i.e., we have that Eqn. (5.191) equals the following:

$$\eta_S^0\langle \dot{t}, \dot{t}\rangle \cdot \eta_S\langle N'\widetilde{N}, c\rangle + \eta_S^0\langle \dot{t}, \dot{f}\rangle \cdot \eta_S\langle N''\widetilde{N}, c\rangle \quad (5.192)$$

Next we turn to the case in which B is neither of the truth constants $\dot{t}$ or $\dot{f}$. In this case we know that both $\eta_S^0\langle B, \dot{t}\rangle = 0$ and $\eta_S^0\langle B, \dot{f}\rangle = 0$ due to Lemma 3.3, Eqn. (3.65), clause (xx), so that the right-hand side of Eqn. (5.187) equals zero in this case, so that Eqn. (5.187) trivially holds.

***In Case of n ⩾ 1*:** In case $n \geqslant 1$ we start with the right-hand side instance of Eqn. (5.187) as follows:

$$\eta_S^{n+1}\langle B, \dot{t}\rangle \cdot \eta_S\langle N'\widetilde{N}, c\rangle + \eta_S^{n+1}\langle B, \dot{f}\rangle \cdot \eta_S\langle N''\widetilde{N}, c\rangle \tag{5.193}$$

Due to Lemma 2.31 on the decomposition of bounded hitting probabilities, Eqn. (2.39), we know that Eqn. (5.193) equals the following:

$$\sum_{B' \in \Lambda} B \to B' \cdot \eta_S^n\langle B', \dot{t}\rangle \cdot \eta_S\langle N'\widetilde{N}, c\rangle + \sum_{B' \in \Lambda} B \to B' \cdot \eta_S^n\langle B', \dot{f}\rangle \cdot \eta_S\langle N''\widetilde{N}, c\rangle \tag{5.194}$$

Eqn. (5.194) can be transformed into

$$\sum_{B' \in \Lambda} B \to B' \cdot (\eta_S^n\langle B', \dot{t}\rangle \cdot \eta_S\langle N'\widetilde{N}, c\rangle + \eta_S^n\langle B', \dot{f}\rangle \cdot \eta_S\langle N''\widetilde{N}, c\rangle) \tag{5.195}$$

Now, the induction hypothesis applies and we have that Eqn. (5.195) is less than or equal ($\leqslant$) to

$$\sum_{B' \in \Lambda} B \to B' \cdot (\mathit{if}(B', N', N'')\widetilde{N} \Rightarrow c) \tag{5.196}$$

Now due to Lemma 3.3, Eqn. (3.65), clause(ix), we have that $B \to B'$ equals $\mathit{if}(B, N', N'') \to \mathit{if}(B', N', N'')$ and therefore, due to Eqn. (3.65), clause(xii), n times, we know that $B \to B'$ equals $\mathit{if}(B, N', N'')\widetilde{N} \to \mathit{if}(B', N', N'')\widetilde{N}$. Therefore, we know that Eqn. (5.196) equals

$$\sum_{B' \in \Lambda} \mathit{if}(B, N', N'')\widetilde{N} \to \mathit{if}(B', N', N'')\widetilde{N} \cdot (\mathit{if}(B', N', N'')\widetilde{N} \Rightarrow c) \tag{5.197}$$

Now, it is possible to rewrite Eqn. (5.197) as follows:

$$\sum_{M \in \{\mathit{if}(B', N', N'')\widetilde{N} \,|\, B' \in \Lambda\}} \mathit{if}(B, N', N'')\widetilde{N} \to M \cdot (M \Rightarrow c) \tag{5.198}$$

Now, as it has fewer summands, we see that Eqn. (5.197) is less than or equal ($\leqslant$) to

$$\sum_{M \in \Lambda} \mathit{if}(B, N', N'')\widetilde{N} \to M \cdot (M \Rightarrow c) \tag{5.199}$$

Due to Eqn. (2.35) we know that Eqn. (5.199) equals

$$\mathit{if}(B, N', N'')\widetilde{N} \Rightarrow c \tag{5.200}$$

Actually, Eqn. (5.200) completes the proof because it equals the left-hand side of Eqn. (5.187). □

Lemma 5.35 (Performance of Conditional Expressions) *Given a performing, Boolean program B and performing, closed terms N' and N'', i.e., $B{:}\mathit{bool}$, $N' : t$ and $N'' : t$ for some type $t \in T$ such that $\alpha(B)$, $\alpha(N')$ and $\alpha(N'')$ we have that $\alpha(\mathit{if}(B, N', N''))$.*

Proof. Given a constant $c : t'$ and closed terms $\widetilde{N} = N_0, \ldots, N_n$ such that $\alpha(N_i)$ for all $0 \leqslant i \leqslant n$ and $\mathit{if}(B, N', N'')\widetilde{N}$ is a program of type t'. We start as follows:

$$[\![\mathit{if}(B, N', N'')\widetilde{N}]\!]_\emptyset[c] \tag{5.201}$$

Due to semantic equation Eqn. (5.32), n times, we have that Eqn. (5.201) equals

$$[\![\mathit{if}(B, N', N'')]\!]_\emptyset\widetilde{[\![N]\!]_\emptyset}[c] \tag{5.202}$$

Due to semantic equation Eqn. (5.30) and the definition of semantic data points in Def. 5.25, i.e., $[\mathit{true}] = \mathcal{T}$ and $[\dot{f}] = \mathcal{F}$, we have that Eqn. (5.202) equals

$$([\![\, B \,]\!]_\emptyset[\dot{t}] \otimes [\![\, N' \,]\!]_\emptyset \oplus [\![\, B \,]\!]_\emptyset[\dot{f}] \otimes [\![\, N'' \,]\!]_\emptyset)\ \widetilde{[\![N]\!]_\emptyset}\ [c] \tag{5.203}$$

Due to the properties of the scalar operations as well as semantic equation Eqn. (5.32), n times, we have that Eqn. (5.203) equals

$$([\![B\,]\!]_\emptyset[\dot{t}] \cdot [\![\, N'\widetilde{N}\,]\!]_\emptyset[c] \;+\; [\![B\,]\!]_\emptyset[\dot{f}] \cdot [\![\, N''\widetilde{N}\,]\!]_\emptyset[c] \tag{5.204}$$

Now, the premises $\alpha(B)$, $\alpha(N')$ and $\alpha(N'')$ apply. Due to $\alpha(B)$ we have that $[\![B]\!]_\emptyset[\dot{t}] \leqslant B \Rightarrow \dot{t}$ and $[\![B]\!]_\emptyset[\dot{f}] \leqslant B \Rightarrow \dot{f}$, because of Def. 5.28, clause (a). Next, due to Def. 5.28, clause (a), n times, the premises $\alpha(N')$ and $\alpha(N'')$, and the fact that $\alpha(N_i)$ for all N_i of $\widetilde{N}$ we have that also $\alpha(N'\widetilde{N})$ and $\alpha(N''\widetilde{N})$. Now, due to $\alpha(N'\widetilde{N})$ we have that $[\![N'\widetilde{N}]\!]_\emptyset[c] \leqslant N'\widetilde{N} \Rightarrow c$. Similarly, we have that $[\![N''\widetilde{N}]\!]_\emptyset[c] \leqslant N''\widetilde{N} \Rightarrow c$. Altogether, we have that Eqn. (5.204) is less than or equal ($\leqslant$) to

$$B \Rightarrow \dot{t} \cdot N'\widetilde{N} \Rightarrow c \;+\; B \Rightarrow \dot{f} \cdot N''\widetilde{N} \Rightarrow c \tag{5.205}$$

Next, due to the big-step approximation Lemma for conditional expressions, i.e., Lemma 5.34, we have that Eqn. (5.205) is less than or equal ($\leqslant$) to

$$\mathit{if}(B, N', N'')\widetilde{N} \Rightarrow c \tag{5.206}$$

Due to the fact that Eqn. (5.201) is less than or equal ($\leqslant$) to Eqn. (5.206) for all constants $c : t'$ and the definition of performing terms Def. 5.28, clause (a), we have that $\alpha(\mathit{if}(B, N', N'')\widetilde{N})$. Therefore, and due to Lemma 5.31, we finally have that $\alpha(\mathit{if}(B, N', N''))$. □

Performance of Applications

We proceed with closed application expressions. This is a particularly easy case in which we do not need a notion of big-step semantics, i.e., this case is an immediate Corollary of Def. 5.28; see Lemma 5.36.

Lemma 5.36 (Performance of Applications) *Given closed performing terms $M : t_1 \to t_2$ and $N : t_1$, i.e., $\alpha(M)$ and $\alpha(N)$, we have that $\alpha(MN)$.*

Proof. The Lemma immediately follows from the definition of performing terms, Def. 5.28. Due to Def. 5.28, clause (b), we have that $\alpha(M)$ and $\alpha(N)$ implies $\alpha(MN)$. □

Performance of Abstractions

We proceed with closed abstraction expressions. Here, we again follow the pattern of establishing a big-step semantics first and the respective performance on the basis of this. We introduce a big-step semantics for abstractions in Lemma 5.37 and use it to prove the performance of abstractions in Lemma 5.38.

Lemma 5.37 (Big-Step Approximation of Abstractions) *Given a closed term $\lambda x.M$, closed terms $N_0, \ldots, N_n$ such that $(\lambda x.M)\widetilde{N}$ is a program and a constant $c \in C$ we have that*

$$((\lambda x.M)N_0N_1 \ldots N_n \Rightarrow c) \geqslant (M[x{:=}N_0]\, N_1 \ldots N_n \Rightarrow c) \tag{5.207}$$

Proof. Henceforth, let us use $\widetilde{N}$ to denote $N_0, N_1, \ldots, N_n$ and $\widetilde{N}'$ to denote $N_1, \ldots, N_n$. We start with the left-hand side of Eqn. (5.207) as follows:

$$(\lambda x.M)\widetilde{N} \Rightarrow c \tag{5.208}$$

Due to the one-step decomposition of hitting probabilities, Eqn. (2.35), we have that Eqn. (5.208) equals

$$\sum_{Q \in \Lambda_\emptyset} (\lambda x.M)\widetilde{N} \to Q \cdot Q \Rightarrow c \tag{5.209}$$

Selecting one particular summand of Eqn. (5.209) we have that Eqn. (5.209) is greater than or equal ($\geqslant$) to

$$(\lambda x.M)\widetilde{N} \to M[x{:=}N_0]\widetilde{N}' \cdot M[x{:=}N_0]\widetilde{N}' \Rightarrow c \tag{5.210}$$

Let us analyze Eqn. (5.210) further and have a look at its sub term $(\lambda x.M)\widetilde{N}$, which actually is $(\lambda x.M)N_0N_1 \ldots N_n$. Due to Lemma 3.3, Eqn. (3.65), (xiii) we have that $(\lambda x.M)N_0 \to M[x := N_0]$ equals one. Now, Lemma 3.3, Eqn. (3.65) (xii), applies n times so that $(\lambda x.M)\widetilde{N} \to M[x{:=}N_0]\widetilde{N}'$ equals one, so that Eqn. (5.210) equals

$$M[x{:=}N_0]\widetilde{N}' \Rightarrow c \tag{5.211}$$

But Eqn. (5.211) is just the right-hand side of Eqn. (5.207) so that it completes the proof. □

Lemma 5.38 (Performance of Abstractions) *Given a performing term M, i.e., $\alpha(M)$, such that $V_{free}(M) = \{x\}$ and $x : t$, we have that $\alpha(\lambda x.M)$.*

Proof. Given a constant $c : t'$ and closed terms $\widetilde{N} = N_0, \ldots, N_n$ such that $\alpha(N_i)$ for all $0 \leqslant i \leqslant n$ and $\lambda x.M\widetilde{N}$ is a program of type t'. We start as follows:

$$[\![\lambda x.M\widetilde{N}]\!]_\emptyset[c] \tag{5.212}$$

Due to semantic equation Eqn. (5.32), n times plus once, we have that Eqn. (5.212) equals

$$[\![\lambda x.M]\!]_\emptyset \; [\![N_0]\!]_\emptyset \; [\![N_1]\!]_\emptyset \cdots [\![N_n]\!]_\emptyset \; [c] \tag{5.213}$$

Due to the semantic equation Eqn. (5.31) we have that Eqn. (5.213) equals

$$(\lambda d \in t \mapsto [\![M]\!]_{\emptyset[x:=d]}) \; [\![N_0]\!]_\emptyset \; [\![N_1]\!]_\emptyset \cdots [\![N_n]\!]_\emptyset \; [c] \tag{5.214}$$

Function application of $(\lambda d \in t \mapsto [\![M]\!]_{\emptyset[x:=d]})$ to $[\![N_0]\!]_\emptyset$ in Eqn. (5.214) yields the following:

$$[\![M]\!]_{\emptyset[x:=[\![N_0]\!]_\emptyset]} \; [\![N_1]\!]_\emptyset \cdots [\![N_n]\!]_\emptyset \; [c] \tag{5.215}$$

Due to the substitution Lemma 5.24 we have that Eqn. (5.215) equals

$$[\![M[x := N_0]]\!]_\emptyset \; [\![N_1]\!]_\emptyset \cdots [\![N_n]\!]_\emptyset \; [c] \tag{5.216}$$

Again due to the semantic equation Eqn. (5.32), n times in this case, we have that Eqn. (5.216) equals

$$[\![M[x := N_0] \, N_1 \ldots N_n]\!]_\emptyset[c] \tag{5.217}$$

Due to the premise $\alpha(M)$ of the Lemma, and the fact that $\alpha(M[x := N_0])$, we have due to Def. 5.28, clause (c) that $\alpha(M[x := N_0] \, N_1 \ldots N_n)$. Therefore, due to Def. 5.28, clause (a) we have that Eqn. (5.217) is less than or equal ($\leqslant$) to the following:

$$M[x := N_0] \, N_1 \ldots N_n \Rightarrow c \tag{5.218}$$

Now, due to the big-step approximation of abstractions, see Lemma 5.37, we have that Eqn. (5.218) is less than or equal ($\leqslant$) to the following:

$$\lambda x.M \; N_0 \; N_1 \ldots N_n \Rightarrow c \tag{5.219}$$

Due to the that fact Eqn. (5.212) is less than or equal ($\leqslant$) to Eqn. (5.219) for all constants $c : t'$ and the definition of performing terms Def. 5.28, clause (a), we have that $\alpha((\lambda x.M)\widetilde{N})$. Therefore, and due to Lemma 5.31, we finally have that $\alpha(\lambda x.M)$. □

Performance of Recursions

We proceed with closed recursion expressions. This is the most elaborate case. We need to introduce a family of standard non-terminating terms. Next, we give some little helper lemmas on the denotational semantics and operational semantics of these non-terminating terms. Next, we need to introduce the syntactical notion of n-fold term application and the syntactical notion of term unwinding. We will establish a big-step semantics for term unwindings and, on the basis of this, will prove the performance of term unwindings. In order to prove the performance of term unwindings we also need a little helper lemma on the semantics of bottom elements in base contexts. Next, we will establish a syntactical notion of approximation, called syntactic recursion approximations. We will investigate how syntactic recursion approximation is preserved under the decomposition of terms. Next, we will introduce a notion of syntactic simulation. All this is needed to understand the semantic correspondence of term unwinding and recursions. Based on that, we will be able to prove the performance of recursions.

Definition 5.39 (Non-Terminating Term) *We define non-terminating terms $\Omega_t : t$ for all types $t \in T$ inductively as follows:*

$$\Omega_{num} = \mu\lambda x_{num} \,.\, x \tag{5.220}$$

$$\Omega_{bool} = \mu\lambda x_{bool} \,.\, x \tag{5.221}$$

$$\Omega_{t_1 \to t_2} = \lambda x_{t_1} \,.\, \Omega_{t_2} \tag{5.222}$$

Lemma 5.40 (Denotational Semantics of Non-Terminating Terms) *For all types $t \in T$ we have*

$$[\![\Omega_t]\!]_\emptyset = \bot_t \tag{5.223}$$

Proof. The proof is straightforward by structural induction over the construction of types $t \in T$.

***In Case of* $\mathbf{t \in T_g}$:** In case of ground types $t \in T_g$ we have that $[\![\Omega_t]\!]_\emptyset$ equals $[\![\mu\lambda x_t \,.\, x]\!]_\emptyset$ due to Def. 5.39, Eqn. (5.220). Due to the semantic equation Eqn. (5.33) and the fixed-point Theorem 2.79 we have that $[\![\mu\lambda x_t \,.\, x]\!]_\emptyset$ equals $\bigsqcup_{n\in\omega} [\![\lambda x_t \,.\, x]\!]^n_\emptyset(\bot_t)$. Now, due to the semantic equation Eqn. (5.32) we know that $[\![\lambda x_t \,.\, x]\!]_\emptyset$ equals $d \in [\![t]\!] \mapsto [\![x]\!]_{\emptyset[x:=d]}$, which equals, due to the semantic equation Eqn. (5.23), the function $d \in [\![t]\!] \mapsto d$, i.e., the identity function $\mathbf{id}_{[\![t]\!]}$ on base domain $[\![t]\!]$. Therefore, we know that also n-fold application $[\![\lambda x_t \,.\, x]\!]^n_\emptyset(\bot_t)$ of $[\![\lambda x_t \,.\, x]\!]_\emptyset$ to $(\bot_t)$ yields $(\bot_t)$ for all $n \in \mathbb{N}_0$. Therefore, we also know that $\bigsqcup_{n\in\omega} [\![\lambda x_t \,.\, x]\!]^n_\emptyset(\bot_t)$ equals $(\bot_t)$.

***In Case of* $\mathbf{t = t_1 \to t_2}$:** In case of higher types $t_1 \to t_2$ we have that $[\![\Omega_{t_1 \to t_2}]\!]_\emptyset$ equals $d \in t_1 \mapsto [\![\Omega_{t_2}]\!]_{\emptyset[x:=d]}$ which equals $d \in t_1 \mapsto [\![\Omega_{t_2}]\!]_\emptyset$, because we have that Ω_{t_2} is a closed term and therefore Lemma 5.23 applies. Now, the induction hypothesis applies and we therefore know that $d \in t_1 \mapsto [\![\Omega_{t_2}]\!]_\emptyset$ equals $d \in t_1 \mapsto \bot_{t_2}$. However, finally we know that by definition of domain bottom elements $d \in t_1 \mapsto \bot_{t_2}$ equals $\bot_{t_1 \to t_2}$; compare, e.g., with Eqn. (5.13). □

Lemma 5.41 (Operational Semantics of Non-Terminating Terms) *For all types $t \in T$ and terms $\widetilde{N} = N_0 \dots N_n$ such that $\Omega_t \widetilde{N}$ has ground type, i.e., there exist $t' \in T_g$ so that $\Omega_t \widetilde{N} : t'$, we have, for all $c \in C_{t'}$ that the following holds:*

$$(\Omega_t \widetilde{N} \Rightarrow c) \;=\; 0 \tag{5.224}$$

Proof. The proof is conducted by structural induction over the construction of types $t \in T$.

***In Case of* $\mathbf{t \in T_g}$:** In case of ground types we know that $\widetilde{N}$ has length $n = 0$, so that prooving $(\Omega_t \widetilde{N} \Rightarrow c) \;=\; 0$ amounts to proving $(\Omega_t \Rightarrow c) \;=\; 0$. Now, in case of ground types $t \in T_g$ the non-terminating term Ω_t is defined as $\mu\lambda x_t \,.\, x$; see Def. 5.39, Eqn. (5.220) – let's drop t from x_t in the sequel for better readability. Due to the definition of termination degree in Def. 4.1 we know that a program M that has a termination degree of zero, i.e., a program M with $\eta_S\langle M, C\rangle = 0$, never hits a constant, which means that $(M \Rightarrow c) = 0$ for all $c \in C$. Therefore, let us prove $(\mu\lambda x.x \Rightarrow c) = 0$ for all $c \in C_t$ by proving that the termination degree $\eta_S\langle \mu\lambda x.x, C\rangle$ equals zero. We follow the algorithm described in Theorem 4.27 to determine the termination degree of $\mu\lambda x.x$. As the first step we use the *cover* algorithm from Theorem 4.26 to determine the finite cover $Cover_S(\mu\lambda x.x)$ of $\mu\lambda x.x$. Actually, we can immediately see that $Cover_S(\mu\lambda x.x)$ consists of the terms $\mu\lambda x.x$ and $\lambda x.x(\mu\lambda x.x)$. Nevertheless, we prove it rigorously on the basis of Theorem 4.26 as an exercise. First, by Eqn. (4.68) we know that $cover(\mu\lambda x.x)$ equals $cover'(\emptyset, \mu\lambda x.x)$. Now, due to Lemma 3.3, Eqn. (3.65), clause (xiv) and clause (xx) we know that $\lambda x.x(\mu\lambda x.x)$ is the one and only term $M \in \Lambda_P$ such that $\mu\lambda x.x \to M > 0$. Therefore, due to the definition of the reduction graph R in Def. 4.14 and its underlying reduction relation ρ as defined in Def. 4.13 we know that the set of outgoing edges of $\mu\lambda x.x$ in E_R consists of $\mu\lambda x.x \to M > 0$ as single element. Therefore, by Eqn. (4.69), third clause, we know that $cover'(\emptyset, \mu\lambda x.x)$ equals

$$cover'(\{\mu\lambda x.x\}, \lambda x.x(\mu\lambda x.x)) \tag{5.225}$$

For a similar line of argumentation, this time based on clause (xiii) and again clause (xx) in Lemma 3.3, Eqn. (3.65), we can see that Eqn. (5.225) equals

$$cover'(\{\mu\lambda x.x, \lambda x.x(\mu\lambda x.x)\}, \mu\lambda x.x) \tag{5.226}$$

However, we have that $\mu\lambda x.x$ belongs to the first argument set of $cover'$ in Eqn. (5.226). Therefore, due to Eqn. (4.69), first clause, we know that $cover'$ terminates as the next step, which finally yields

$$Covers_S(\mu\lambda x.x) = \{\mu\lambda x.x, \lambda x.x(\mu\lambda x.x)\} \tag{5.227}$$

Now, due to Theorem 4.27, Eqn. (4.85), and again Lemma 3.3, Eqn. (3.65), clauses (xiv), (xiii) and (xx), we have that the vector of hitting probabilities consisting of $\eta_S\langle\mu\lambda x.x, C\rangle$ and $\eta_S\langle\lambda x.x(\mu\lambda x.x), C\rangle$ is determined as the least solution of the following finite linear equation system:

$$\begin{aligned} \eta_S\langle\mu\lambda x.x, C\rangle &= \eta_S\langle\lambda x.x(\mu\lambda x.x), C\rangle \\ \eta_S\langle\lambda x.x(\mu\lambda x.x), C\rangle &= \eta_S\langle\mu\lambda x.x, C\rangle \end{aligned} \tag{5.228}$$

Finally, the least solution Eqn. (5.228) yields that $\mu\lambda x.x$ equals zero.

***In Case of* $\mathbf{t = t_1 \to t_2}$:** In case of higher types we know due to Def. 5.39, Eqn. (5.222) that $\Omega_{t_1\to t_2}$ equals $\lambda x_{t_1}.\Omega_{t_2}$. Due to Eqn. (2.35) we have that $\Omega_{t_1\to t_2}\widetilde{N} \Rightarrow c$ equals

$$\sum_{M\in\Lambda_\emptyset} ((\lambda x_{t_1}.\Omega_{t_2})\widetilde{N} \to M) \cdot (M \Rightarrow c) \tag{5.229}$$

Remember that $\widetilde{N} = N_0N_1 \ldots N_n$. In the next few equations, let $\widetilde{N'}$ denote the tail of $\widetilde{N}$ after dropping N_0, i.e., let $\widetilde{N'} = N_1 \ldots N_n$. Now, we can always transform Eqn. (5.229) into the following:

$$((\lambda x_{t_1}.\Omega_{t_2})\widetilde{N} \to \Omega_{t_2}\widetilde{N'}) \cdot (\Omega_{t_2}\widetilde{N'} \Rightarrow c) + \sum_{M\in\Lambda_\emptyset\setminus\{\Omega_{t_2}\widetilde{N'}\}} ((\lambda x_{t_1}.\Omega_{t_2})\widetilde{N} \to M) \cdot (M \Rightarrow c) \tag{5.230}$$

With respect to the second summand in Eqn. (5.230), we know, due to Lemma 3.3, Eqn. (3.65), clause (xx) that $(\lambda x_{t_1}.\Omega_{t_2})\widetilde{N} \to M$ equals zero for all $M \neq \Omega_{t_2}\widetilde{N'}$. With respect to the first summand in Eqn. (5.230) we can apply the induction hypothesis and know that $\Omega_{t_2}\widetilde{N'} \Rightarrow c$ equals zero. Therefore, overall we have that Eqn. (5.230) equals zero, which completes the proof. □

Definition 5.42 (n-fold Term Application) *For each term of some higher type $M : t \to t$ we define a term $M^n : t \to t$, called the n-fold application of term M, for all $n \in \mathbb{N}_0$ and arbitrary terms $N : t$ of appropriate type t inductively as follows:*

$$M^0N = N \tag{5.231}$$

$$M^{n+1}N = M(M^nN) \tag{5.232}$$

Definition 5.43 (Unwinding Term) *For each term of some higher type $M : t \to t$ we define a term $\langle M^n\Omega\rangle : t \to t$, called the n-times unwinding of term M, or unwinding term for short, for all $n \in \mathbb{N}_0$ inductively as follows:*

$$\langle M^0\Omega\rangle = \Omega_{t\to t} \tag{5.233}$$

$$\langle M^{n+1}\Omega\rangle = (\lambda f_{t\to t} \cdot (f\ \langle M^n\Omega\rangle))\ M \tag{5.234}$$

Lemma 5.44 (Semantics of Bottom Elements in Base Contexts) *For all types $t \in T$ and data $\widetilde{d} = d_0 \dots d_n$ such that $\bot_t \widetilde{d}$ is a base element, i.e., there exists $t' \in T_g$ so that $\bot_t \widetilde{d} \in [\![t']\!]$ with $[\![t']\!] = \lfloor S \longrightarrow [0,1]_\omega \rfloor$ for some set S, we have, for all data points $p \in S$ that the following holds:*

$$\bot_t \widetilde{d} p \;=\; 0 \tag{5.235}$$

Proof. This lemma follows as an immediate corollary from the definitions of the established domains in Sect. 5.1.1. We have that $\bot_t \widetilde{d}$ equals $\bot_{t'}$ due to Eqn. (5.13), n times, and therefore $\bot_t \widetilde{d}p = 0$ due to Eqn. (5.7) resp. Eqn. (5.10) one time. □

Lemma 5.45 (Big-Step Approximation of Term Unwindings) *Given a closed term $M : t \to t$, closed terms $\widetilde{N} = N_1, \dots, N_n$ such that $\langle M^{n+1}\Omega\rangle\widetilde{N}$ is a program and a constant $c \in C$ we have that*

$$\langle M^{n+1}\Omega\rangle\widetilde{N} \Rightarrow c \;\geqslant\; (M(\langle M^n\Omega\rangle\widetilde{N}) \Rightarrow c) \tag{5.236}$$

Proof. Due to the one-step decomposition of hitting probabilities, Eqn. (2.35), we have that the left-hand side of Eqn. (5.236) equals the following:

$$\sum_{Q \in \Lambda_\emptyset} \langle M^{n+1}\Omega\rangle\widetilde{N} \to Q \,\cdot\, Q \Rightarrow c \tag{5.237}$$

Now, we can select one particular summand of Eqn. (5.237) and have that Eqn. (5.237) is greater than or equal ($\geqslant$) to the following:

$$\langle M^{n+1}\Omega\rangle\widetilde{N} \to M(\langle M^n\Omega\rangle\widetilde{N}) \,\cdot\, M(\langle M^n\Omega\rangle\widetilde{N}) \Rightarrow c \tag{5.238}$$

Let us consider the following sub-term of Eqn. (5.238), i.e.,

$$\langle M^{n+1}\Omega\rangle \tag{5.239}$$

Now, due to the definition of unwinding terms, Def. 5.43, Eqn. (5.234), we have that Eqn. (5.239) is identical to the following:

$$(\lambda f_{t\to t} \,.\, (f\,\langle M^n\Omega\rangle))M \tag{5.240}$$

Due to Lemma 3.3, Eqn. (3.65), clause (xiii), we know the following:

$$((\lambda f_{t\to t} \,.\, (f\,\langle M^n\Omega\rangle))M \to ((f\,\langle M^n\Omega\rangle)[f := M])) \;=\; 1 \tag{5.241}$$

Now, due to the definition of substitution, the fact that M is a closed term, and again Def. 5.43, Eqn. (5.234), we know that Eqn. (5.241) is equivalent to the following:

$$(\langle M^{n+1}\Omega\rangle \to (M\,\langle M^n\Omega\rangle)) \;=\; 1 \tag{5.242}$$

Then, due to Lemma 3.3, Eqn. (3.65), clause (xii), applied n times, we have that Eqn. (5.242) implies

$$(\langle M^{n+1}\Omega\rangle\widetilde{N} \rightarrow (M\,\langle M^{n}\Omega\rangle)\widetilde{N}) \;=\; 1 \tag{5.243}$$

Now, Eqn. (5.243) just states that the left factor of Eqn. (5.238) equals one. Therefore, we have that Eqn. (5.238) equals

$$M(\langle M^{n}\Omega\rangle\widetilde{N}) \Rightarrow c \tag{5.244}$$

Now, with Eqn. (5.244) we are finished, because it is the right-hand side of Eqn. (5.236). □

Lemma 5.46 (Performance of Unwinding Terms) *Given a performing, closed term $M : t \rightarrow t$ we have that $\alpha(\langle M^{n}\Omega\rangle)$ for all $n \in \mathbb{N}_0$.*

Proof. Given a constant $c : t'$ and closed terms $\widetilde{N} = N_0, \ldots, N_n$ such that $\alpha(N_i)$ for all $0 \leqslant i \leqslant n$ and $\langle M^{n}\Omega\rangle\widetilde{N}$. We proceed with showing that $\alpha(\langle M^{n}\Omega\rangle\widetilde{N})$ by a natural induction over the number of unwindings n. From this we will then conclude that $\alpha(\langle M^{n}\Omega\rangle)$.

***In Case of n=0*:** In the base case, we start as follows:

$$[\![\langle M^{0}\Omega\rangle\widetilde{N}]\!]_{\emptyset}[c] \tag{5.245}$$

Due to the semantic equation Eqn. (5.32), n times, we have that Eqn. (5.245) equals

$$[\![\langle M^{0}\Omega\rangle]\!]_{\emptyset}\widetilde{[\![N]\!]_{\emptyset}}[c] \tag{5.246}$$

Due to the definition of unwinding terms, compare with Eqn. (5.233), we have that Eqn. (5.246) equals

$$[\![\Omega_{t\rightarrow t}]\!]_{\emptyset}\widetilde{[\![N]\!]_{\emptyset}}[c] \tag{5.247}$$

Due to the denotational semantics of non-terminating terms, i.e., Lemma 5.40, we have that Eqn. (5.247) equals

$$[\![\bot_{t\rightarrow t}]\!]_{\emptyset}\widetilde{[\![N]\!]_{\emptyset}}[c] \tag{5.248}$$

Due to the denotational semantics of bottom elements in base contexts, i.e., Lemma 5.44, we have that Eqn. (5.248) equals zero for all constants and therefore we have that $\alpha(\langle M^{0}\Omega\rangle\widetilde{N})$.

***In Case of $n \geqslant 1$*:** In case the number of unwindings is $n \geqslant 1$, we start as follows:

$$[\![\langle M^{n+1}\Omega\rangle\widetilde{N}]\!]_{\emptyset}[c] \tag{5.249}$$

Again, due to the semantic equation Eqn. (5.32), n times, we have that Eqn. (5.249) equals

$$[\![\langle M^{n+1}\Omega\rangle]\!]_{\emptyset}\widetilde{[\![N]\!]_{\emptyset}}[c] \tag{5.250}$$

Due to the definition of unwinding terms, compare with Eqn. (5.234), we have that Eqn. (5.250) equals

$$[\![(\lambda f_{t\to t}\,.\,(f\,\langle M^n\Omega\rangle))\;M]\!]_\emptyset\widetilde{[\![N]\!]_\emptyset}[c] \tag{5.251}$$

Due to semantic equation Eqn. (5.32) we have that Eqn. (5.251) equals

$$[\![(\lambda f_{t\to t}\,.\,(f\,\langle M^n\Omega\rangle))]\!]_\emptyset\;[\![M]\!]_\emptyset\widetilde{[\![N]\!]_\emptyset}[c] \tag{5.252}$$

Due to semantic equation Eqn. (5.31) we have that Eqn. (5.252) equals

$$d\in[\![t\to t]\!]\mapsto[\![f\,\langle M^n\Omega\rangle]\!]_{\emptyset[x:=d]}\;[\![M]\!]_\emptyset\widetilde{[\![N]\!]_\emptyset}[c] \tag{5.253}$$

After function application we have that Eqn. (5.253) yields

$$[\![f\,\langle M^n\Omega\rangle]\!]_{\emptyset[f:=[\![M]\!]_\emptyset]}\;\widetilde{[\![N]\!]_\emptyset}[c] \tag{5.254}$$

Due to the substitution Lemma 5.24 we have that Eqn. (5.254) equals

$$[\![(f\,\langle M^n\Omega\rangle)[x:=M]]\!]_\emptyset\;\widetilde{[\![N]\!]_\emptyset}[c] \tag{5.255}$$

Due to the definition of substitution plus the fact that M is a closed term we have that Eqn. (5.255) equals

$$[\![M(\langle M^n\Omega\rangle)]\!]_\emptyset\;\widetilde{[\![N]\!]_\emptyset}[c] \tag{5.256}$$

Due to semantic equation Eqn. (5.32) we have that Eqn. (5.256) equals

$$[\![(M(\langle M^n\Omega\rangle))\;\widetilde{N}]\!]_\emptyset[c] \tag{5.257}$$

Now, we can apply the induction hypothesis $\alpha(\langle M^n\Omega\rangle\widetilde{N})$. Due to $\alpha(\langle M^n\Omega\rangle\widetilde{N})$ and Lemma 5.31 we also have that $\alpha(\langle M^n\Omega\rangle)$. Furthermore, we can exploit $\alpha(M)$, which is a premise of the lemma. Due to $\alpha(M)$, $\alpha(\langle M^n\Omega\rangle)$, and Def. 5.28, clause (b), we have that $\alpha(M(\langle M^n\Omega\rangle))$. Now, due to $\alpha(M(\langle M^n\Omega\rangle))$ and again Def. 5.28, clause (b), n times, we have that $\alpha((M(\langle M^n\Omega\rangle))\,\widetilde{N})$. Based on $\alpha((M(\langle M^n\Omega\rangle))\,\widetilde{N})$ and Def. 5.28, clause (a) we are able to see that Eqn. (5.257) is less than or equal ($\leqslant$) to the following:

$$(M(\langle M^n\Omega\rangle))\;\widetilde{N}\Rightarrow c \tag{5.258}$$

Next, we can apply the big-step approximation of term unwindings, i.e., Lemma 5.45, to Eqn. (5.258), so that we know that Eqn. (5.258) is less than or equal ($\leqslant$) to the following:

$$(\langle M^{n+1}\Omega\rangle)\;\widetilde{N}\Rightarrow c \tag{5.259}$$

Together with Def. 5.28, clause (a), we have that Eqn. (5.259) amounts to $\alpha(\langle M^{n+1}\Omega\rangle\widetilde{N})$ which completes our natural induction proof of $\alpha(\langle M^n\Omega\rangle\widetilde{N})$ for all $n\in\mathbb{N}_0$. Now, as the final step we have, due to $\alpha(\langle M^n\Omega\rangle\widetilde{N})$ and Lemma 5.31, that $\alpha(\langle M^n\Omega\rangle)$. □

Definition 5.47 (Syntactic Recursion Approximation) *We define the syntactic approximation relation* $\preccurlyeq : \Lambda \times \Lambda$ *inductively by the following set of rules:*

$$(i)\frac{M : t}{\Omega_t \preccurlyeq M} \qquad (ii)\frac{M : t \to t \quad n \in \mathbb{N}_0}{\langle M^n \Omega \rangle \preccurlyeq \mu M}$$

$$(iii)\frac{}{M \preccurlyeq M} \qquad (iv)\frac{M \preccurlyeq M'}{\lambda x.M \preccurlyeq \lambda x.M'} \qquad (v)\frac{M \preccurlyeq M' \quad N \preccurlyeq N'}{MN \preccurlyeq M'N'}$$

$$(vi)\frac{M \preccurlyeq M' \quad N \preccurlyeq N'}{M|N \preccurlyeq M'|N'}$$

Lemma 5.48 proves that the syntactic approximation is preserved under the one-step semantics. First, Lemma 5.48 refines Lemma 3.3. Lemma 3.3 states that the one-step transition is a total function; now, Lemma 5.48 makes explicit why in terms of the concrete possible next reduction steps that can be made for a term. On the basis of this, Lemma 5.48 states the preservation property that is also illustrated in Fig. 5.1.

Lemma 5.48 (Preservation of Syntactic Approximation) *Given a closed term* P*. Then (i)* **either** *there exists a closed term* P' *such that* $P \xrightarrow{1} P'$ *and* $P \xrightarrow{0} P''$ *for all* $P'' \neq P'$ **or** *there exist closed terms* P' *and* P'' *such that* $P' \neq P''$, $P \xrightarrow{0.5} P'$, $P \xrightarrow{0.5} P''$ *and* $P \xrightarrow{0} P'''$ *for all* $P''' \neq P'$ *and* $P''' \neq P''$*. Furthermore, (ii) given a closed term* Q *such that* $P \preccurlyeq Q$*. Then, in case there exists a* P' *with* $P \xrightarrow{1} P'$ *there exists a* Q' *with* $Q \xrightarrow{1} Q'$ *such that* $P' \preccurlyeq Q'$*. Furthermore, in case there exist closed terms* P' *and* P'' *with* $P' \neq P''$, $P \xrightarrow{0.5} P'$ *and* $P \xrightarrow{0.5} P''$ *there exist closed terms* Q' *and* Q'' *with* $Q' \neq Q''$, $P \xrightarrow{0.5} Q'$ *and* $Q \xrightarrow{0.5} Q''$ *such that* $P' \preccurlyeq Q'$ *and* $P'' \preccurlyeq Q''$*.*

Proof. The Lemma can be proven by structural induction over the construction of term P on the basis of Lemma 3.3. □

$$(i)\;\begin{array}{ccc} Q & \xrightarrow{1} & Q' \\ \curlyvee & & \curlyvee \\ \boldsymbol{P} & \xrightarrow{1} & \boldsymbol{P'} \end{array} \qquad\qquad (ii)\;\begin{array}{ccccc} Q' & \xleftarrow{0.5} & Q & \xrightarrow{0.5} & Q'' \\ \curlyvee & & \curlyvee & & \curlyvee \\ \boldsymbol{P'} & \xleftarrow{0.5} & \boldsymbol{P} & \xrightarrow{0.5} & \boldsymbol{P''} \end{array}$$

Fig. 5.1. Preservation of syntactic approximation under reduction

Lemma 5.49 (Syntactic Simulation) *Given closed terms* A *and* B *such that* $A \preccurlyeq B$ *and closed terms* $\widetilde{N} = N_0, \ldots, N_n$ *such that* $A\widetilde{N}$ *and* $B\widetilde{N}$ *are programs we have that*

$$A\widetilde{N} \Rightarrow c \leqslant B\widetilde{N} \Rightarrow c \qquad (5.260)$$

Proof. Due to the definition of reduction probabilites in Def. 3.5 and Corollary 2.27 we have that Eqn. (5.260) is equivalent to

$$\lim_{n\to\infty} \eta_S^n\langle A\widetilde{N}, c\rangle \leqslant \lim_{n\to\infty} \eta_S^n\langle B\widetilde{N}, c\rangle \tag{5.261}$$

Due to the monotonicity of bounded hitting probabilities, i.e., Lemma 2.28 and the monotone convergence Theorem 2.93, we have that both of the limits in Eqn. (5.166) are also the least upper bound of their respective chains. Therefore, in order to prove Eqn. (5.261) it suffices to prove that the following holds for all $n \in \mathbb{N}_0$:

$$\eta_S^n\langle A\widetilde{N}, c\rangle \leqslant \eta_S^n\langle B\widetilde{N}, c\rangle \tag{5.262}$$

We can prove Eqn. (5.262) by a natural induction over the bound $n \in \mathbb{N}_0$. We show the base case $n = 0$ only. In case of $\eta_S^0\langle A\widetilde{N}, c\rangle$ it suffices to consider two general cases, i.e., the case that $A\widetilde{N} \neq c$ and the case that $A\widetilde{N} = c$. In the case that $A\widetilde{N} \neq c$ we have, due to the definition of bounded hitting probabilities, Def. 2.24, that $\eta_S^0\langle A\widetilde{N}, c\rangle$ equals zero, and therefore $\eta_S^0\langle A\widetilde{N}, c\rangle$ is less than or equal to $P \Rightarrow c$ for all programs P and, as a special case, also for all programs B with $A \preccurlyeq B$. In the case that $A\widetilde{N} = c$ we have that $A\widetilde{N}$ is a constant, just because $c \in C$. As a matter of detail, we know that $\widetilde{N}$ is empty in this case with $n = 0$ and $A \in C$. Now, in case A is a constant we know, due to the definition of syntactic approximation in Def. 5.47, that A is the only term that is syntactically approximated by A, and we have that there is no B with $A \neq B$ and $A \preccurlyeq B$. This is so, because syntactic approximation is inductively defined by Def. 5.47 and is therefore the least relation closed under the rules in Def. 5.47. Now, rule (iii) in Def. 5.47, which is the rule of reflexivity, is the only one that is applicable to deduce a term B such that $A \preccurlyeq B$ in case that A is a constant. □

Lemma 5.50 (Semantic Equality of Unwindings and Recursions) *Given a closed term M and an arbitrary number $n \in \mathbb{N}_0$ we have that*

$$[\![\mu M]\!]_\emptyset = \bigsqcup_{n\in\omega} [\![\langle M^n \Omega\rangle]\!]_\emptyset \tag{5.263}$$

Proof. Due to semantic equation Eqn. (5.31) and the fixed-point Theorem 2.79 we know that $[\![\mu M]\!]_\emptyset$ equals

$$\bigsqcup_{n\in\omega} [\![M]\!]_\emptyset^n(\bot_{t\to t}) \tag{5.264}$$

Therefore, in order to prove Eqn. (5.263) it suffices to show that the following holds for all numbers $n \in \mathbb{N}_0$ of unwindings:

$$[\![\langle M^n \Omega\rangle]\!]_\emptyset = [\![M]\!]_\emptyset^n(\bot_{t\to t}) \tag{5.265}$$

We prove Eqn. (5.265) by natural induction over n.

In Case of $n = 0$**:** Due to Def. 5.43 we have that $[\![\langle M^0\Omega\rangle]\!]_\emptyset$ equals $[\![\Omega_{t\to t}]\!]_\emptyset$. Due to Lemma 5.40 we have that $[\![\Omega_{t\to t}]\!]_\emptyset$ equals $\perp_{t\to t}$. Finally, we immediately know that $\perp_{t\to t}$ equals $[\![M]\!]^0_\emptyset(\perp_{t\to t})$.

In Case of $n > 0$**:** Due to the definition of unwinding terms in Def. 5.43 we have that $[\![\langle M^{n+1}\Omega\rangle]\!]_\emptyset$ equals

$$[\![(\lambda f_{t\to t}\,.\,(f\,\langle M^n\Omega\rangle))\;M]\!]_\emptyset \tag{5.266}$$

Due to the semantic equation Eqn. (5.32) we have that Eqn. (5.266) equals

$$[\![\lambda f_{t\to t}\,.\,(f\,\langle M^n\Omega\rangle)]\!]_\emptyset\;[\![M]\!]_\emptyset \tag{5.267}$$

Due to the semantic equation Eqn. (5.31) we have that Eqn. (5.267) equals

$$(d\in[\![t\to t]\!]\mapsto[\![f\,\langle M^n\Omega\rangle]\!]_{\emptyset[f:=d]})\;[\![M]\!]_\emptyset \tag{5.268}$$

Due to ordinary function application we have that Eqn. (5.268) equals

$$[\![f\,\langle M^n\Omega\rangle]\!]_{\emptyset[f:=[\![M]\!]_\emptyset]} \tag{5.269}$$

Again, due to the semantic equation Eqn. (5.32) we have that Eqn. (5.269) equals

$$[\![f]\!]_{\emptyset[f:=[\![M]\!]_\emptyset]}\;[\![\langle M^n\Omega\rangle]\!]_{\emptyset[f:=[\![M]\!]_\emptyset]} \tag{5.270}$$

Due to the semantic equation Eqn. (5.23) applied to the function Eqn. (5.270) plus the fact that the $\langle M^n\Omega\rangle$ in Eqn. (5.270) is a closed term and again semantic equation Eqn. (5.23), we have that Eqn. (5.270) equals

$$[\![M]\!]_\emptyset\;[\![\langle M^n\Omega\rangle]\!]_\emptyset \tag{5.271}$$

Now, it is possible to apply the induction hypothesis to $\langle M^n\Omega\rangle$. Therefore we know that Eqn. (5.271) equals

$$[\![M]\!]_\emptyset\;([\![M]\!]^n_\emptyset(\perp_{t\to t})) \tag{5.272}$$

Actually, with Eqn. (5.272) the proof is completed, because Eqn. (5.272) simply equals $[\![M]\!]^{n+1}_\emptyset(\perp_{t\to t})$; see Def. 2.83 for n-fold function application.

□

Lemma 5.51 (Performance of Recursions) *Given a performing, closed term M, i.e., $\alpha(M)$, we have that $\alpha(\mu M)$.*

Proof. Given a constant $c : t'$ and closed terms $\widetilde{N} = N_1,\ldots,N_n$ such that $\alpha(N_i)$ for all $0 \leqslant i \leqslant n$ and $\mu M\widetilde{N}$ is a program of type t'. We start as follows:

$$[\![(\mu M)\widetilde{N}]\!]_\emptyset[c] \tag{5.273}$$

Due to semantic equation Eqn. (5.32), n times, we have that Eqn. (5.273) equals

$$[\![\mu M]\!]_{\emptyset} \widetilde{[\![N]\!]_{\emptyset}}[c] \tag{5.274}$$

Due to Lemma 5.50 we have that Eqn. (5.274) equals

$$\Big(\bigsqcup_{n\in\omega}[\![\langle M^n\Omega\rangle]\!]_{\emptyset}\Big)\ \widetilde{[\![N]\!]_{\emptyset}}\ [c] \tag{5.275}$$

Due to the definition of least upper bounds of chains of functions that have an ω-cpo as range, i.e., Lemma 2.64, n times, we have that Eqn. (5.275) equals the following:

$$\Big(\bigsqcup_{n\in\omega}\big([\![\langle M^n\Omega\rangle]\!]_{\emptyset}\ \widetilde{[\![N]\!]_{\emptyset}}\big)\Big)\ [c] \tag{5.276}$$

Due to semantic equation Eqn. (5.32), n times, we have that Eqn. (5.276) equals

$$\Big(\bigsqcup_{n\in\omega}[\![\langle M^n\Omega\rangle\ \widetilde{N}]\!]_{\emptyset}\Big)\ [c] \tag{5.277}$$

Again due to Def. 2.64 we know that Eqn. (5.277) equals

$$\bigsqcup_{n\in\omega}\big([\![\langle M^n\Omega\rangle\ \widetilde{N}]\!]_{\emptyset}\ [c]\big) \tag{5.278}$$

Next, we turn our attention to the single elements of the chain in Eqn. (5.278), i.e., $[\![\langle M^n\Omega\rangle \widetilde{N}]\!]_{\emptyset}$ for all numbers $n \in \mathbb{N}_0$. The idea is to show that $\mu M \widetilde{N} \Rightarrow c$ is an upper bound of all $[\![\langle M^n\Omega\rangle\ \widetilde{N}[c]]\!]_{\emptyset}$ and therefore also greater than or equal to the least upper bound in Eqn. (5.278). Now, we have $\alpha(M)$ as a premise of this lemma. With $\alpha(M)$ we can apply Lemma 5.46, which ensures the performance of unwinding terms, so that we have $\alpha(\langle M^n\Omega\rangle)$ for all $n \in \mathbb{N}_0$. Now that we have $\alpha(\langle M^n\Omega\rangle)$ and due to the fact that $\alpha(N_i)$ for all N_i of $\widetilde{N}$ we also have that $\alpha(\langle M^n\Omega\rangle\ \widetilde{N})$ due to the definition of performing terms Def. 5.28, clause (b), n times. Based on $\alpha(\langle M^n\Omega\rangle \widetilde{N})$ and Def. 5.28, clause (a), we can conclude that the following holds:

$$[\![\langle M^n\Omega\rangle\ \widetilde{N}]\!]_{\emptyset}[c] \leqslant \langle M^n\Omega\rangle\ \widetilde{N} \Rightarrow c \tag{5.279}$$

Next, by the definition of unwinding term approximation in Def. 5.47, rule (*ii*) once and rule (*v*) n times, we know the following:

$$\langle M^n\Omega\rangle\ \widetilde{N} \preccurlyeq \mu M\ \widetilde{N} \tag{5.280}$$

Now, due to the fact that $\langle M^n\Omega\rangle$ syntactically approximates $\mu M\ \widetilde{N}$, i.e., Eqn. (5.280), we can apply the syntactic simulation Lemma 5.49 so that we have the following:

$$\langle M^n\Omega\rangle\ \widetilde{N} \Rightarrow c \leqslant \mu M\ \widetilde{N} \Rightarrow c \tag{5.281}$$

From Eqn. (5.279) and Eqn. (5.281) we can conclude that for all $n \in \mathbb{N}_0$ we have that

$$[\![\langle M^n\Omega\rangle\,\widetilde{N}]\!]_\emptyset[c] \leqslant \mu M\,\widetilde{N} \Rightarrow c \tag{5.282}$$

With Eqn. (5.282) we know that $(\mu M)\,\widetilde{N} \Rightarrow c$ is an upper bound for all chain elements in Eqn. (5.278) and therefore is also greater than their least upper bound Eqn. (5.278). Remember that we have already shown that Eqn. (5.273) equals Eqn. (5.278), which almost finishes the proof, i.e., we have that

$$[\![(\mu M)\widetilde{N}]\!]_\emptyset[c] \leqslant \mu M\,\widetilde{N} \Rightarrow c \tag{5.283}$$

By Def. 5.28, clause (a), we have that Eqn. (5.283) means $\alpha(\mu M\widetilde{N})$. Finally, with $\alpha(\mu M\widetilde{N})$ and Def. 5.28, clause (b), n times, we have that $\alpha(\mu M)$. □

Performance of Choice Expressions

The case of choice expressions is again straightforward. First, we introduce a big-step semantics for choices in Lemma 5.52, then we prove the performance of choices in Lemma 5.53.

Lemma 5.52 (Big-Step Approximation of Choices) *Given closed terms M_1 and M_2, closed terms $N_1,\ldots,N_n$ such that $(M_1|M_2)\widetilde{N}$ is a program and a constant $c \in C$ we have that*

$$((M_1|M_2)\widetilde{N}\Rightarrow c) \geqslant (0.5\cdot M_1\widetilde{N}\Rightarrow c + 0.5\cdot M_2\widetilde{N}\Rightarrow c) \tag{5.284}$$

Proof. Due to the one-step decomposition of hitting probabilities, Eqn. (2.35), we have that the left-hand side of Eqn. (5.284), i.e., $(M_1|M_2)\widetilde{N}$ equals

$$\sum_{Q\in\Lambda_\emptyset} (M_1|M_2)\widetilde{N} \rightarrow Q\cdot Q \Rightarrow c \tag{5.285}$$

Let us assume $M_1 \neq M_2$. Now, we can select two particular summands of Eqn. (5.285) and have that Eqn. (5.285) is greater than or equal ($\geqslant$) to the following:

$$((M_1|M_2)\widetilde{N} \rightarrow M_1\widetilde{N}\cdot M_1\widetilde{N} \Rightarrow c)+((M_1|M_2)\widetilde{N} \rightarrow M_2\widetilde{N}\cdot M_2\widetilde{N} \Rightarrow c) \tag{5.286}$$

Due to Lemma 3.3, Eqn. (3.65), clause (xv), we know that $(M_1|M_2)\rightarrow M_1$ equals 0.5. Then, due to Lemma 3.3, Eqn. (3.65), clause (xii), applied n times, we have that $(M_1|M_2)\widetilde{N}\rightarrow M_1\widetilde{N}$ equals 0.5. Analogously, due to Lemma 3.3, Eqn. (3.65), clause (xvi) we have that $(M_1|M_2)\rightarrow M_2$ equals 0.5 and again, due to Lemma 3.3, Eqn. (3.65), (xii), n times, that also $(M_1|M_2)\widetilde{N}\rightarrow M_2\widetilde{N}$ equals 0.5. Altogether, we therefore have that Eqn. (5.286) equals

$$0.5\cdot M_1\widetilde{N}\Rightarrow c + 0.5\cdot M_2\widetilde{N}\Rightarrow c \tag{5.287}$$

Eqn. (5.287) is the right-hand side of Eqn. (5.284) and therefore completes the proof for the case $M_1 \neq M_2$. In case $M_1 = M_2$ the proof is similar by exploiting Eqn. (3.65), clause (xvii). □

Lemma 5.53 (Performance of Choices) *Given performing, closed terms M_1 and M_2, i.e., $\alpha(M_1)$ and $\alpha(M_2)$, we have that $\alpha(M_1|M_2)$.*

Proof. Given a constant $c : t'$ and closed terms $\widetilde{N} = N_0, \ldots, N_n$ such that $\alpha(N_i)$ for all $0 \leqslant i \leqslant n$ and $(M_1|M_2)\widetilde{N}$ is a program of type t'. We start as follows:

$$[\![(M_1|M_2)\widetilde{N}]\!]_\emptyset[c] \tag{5.288}$$

Due to semantic equation Eqn. (5.32), n times plus once, we have that Eqn. (5.288) equals

$$[\![M_1|M_2]\!]_\emptyset \; \widetilde{[\![N]\!]_\emptyset} \; [c] \tag{5.289}$$

Due to semantic equation Eqn. (5.34) we have that Eqn. (5.289) equals

$$(0.5 \otimes [\![M_1]\!]_\emptyset \oplus 0.5 \otimes [\![M_2]\!]_\emptyset) \; \widetilde{[\![N]\!]_\emptyset} \; [c] \tag{5.290}$$

Due to the properties of the scalar operations as well as semantic equation Eqn. (5.32), n times, we have that Eqn. (5.290) equals

$$0.5 \cdot [\![M_1\widetilde{N}]\!]_\emptyset[c] \; + \; 0.5 \cdot [\![M_2\widetilde{N}]\!]_\emptyset[c] \tag{5.291}$$

Now, due to Def. 5.28, clause (a), n times, the premises $\alpha(M_1)$ and $\alpha(M_2)$ and the fact that $\alpha(N_i)$ for all N_i of $\widetilde{N}$ we have that also $\alpha(M_1\widetilde{N})$ and $\alpha(M_2\widetilde{N})$. Due to $\alpha(M_1\widetilde{N})$, we have that $[\![M_1\widetilde{N}]\!]_\emptyset[c] \leqslant M_1\widetilde{N} \Rightarrow c$. Due to $\alpha(M_2\widetilde{N})$, we have that $[\![M_2\widetilde{N}]\!]_\emptyset[c] \leqslant M_2\widetilde{N} \Rightarrow c$. Altogether, we have that Eqn. (5.291) is less than or equal ($\leqslant$) to

$$0.5 \cdot M_1\widetilde{N} \Rightarrow c \; + \; 0.5 \cdot M_2\widetilde{N} \Rightarrow c \tag{5.292}$$

Next, due to the big-step approximation lemma for choices, i.e., Lemma 5.52, we have that Eqn. (5.292) is less than or equal ($\leqslant$) to

$$(M_1|M_2)\widetilde{N} \Rightarrow c \tag{5.293}$$

Due to the fact that Eqn. (5.288) is less than or equal ($\leqslant$) to Eqn. (5.293) for all constants $c : t'$ and the definition of performing terms Def. 5.28, clause (a), we have that $\alpha((M_1|M_2)\widetilde{N})$. Therefore, and due to Lemma 5.31, we finally have that $\alpha(M_1|M_2)$. □

Complete Semantic Correspondence

Lower bound correspondence means that the denotational semantics approximates the Markov chain semantics from below. It means that the given denotational semantics yields a lower bound for the intended denotational semantics that we expect to equal the evaluation semantics. On the basis of the results achieved so far, we now prove the lower bound correspondence and, finally, the complete semantic correspondence between the denotational and the operational semantics.

Lemma 5.54 (Lower Bound Correspondence) *The denotational semantics approximates the evaluation semantics from below. For each program M and constant c we have that*

$$(M \Rightarrow c) \geqslant [\![M]\!]_{\emptyset}([c]) \tag{5.294}$$

Proof: Now, we prove by structural induction that all terms are performing, i.e., $\alpha(M)$ for all $M \in \Lambda$. Proving that all terms are performing implies the desired Lemma 5.54, because Lemma 5.54 is included as a special case in the definition of performing terms Def. 5.28, i.e., as clause (a). We show the cases of variables, constants and basic operators.

In Case of Variables: Given a variable x. According to clause (c) of Def. 5.28 we need to show that $\alpha(M)$ implies $\alpha(x[x := M])$ for each closed lambda term M of appropriate type. This follows immediately, because $x[x := M] = M$.

In Case of Constants: Constants are programs and therefore clause (a) of Def. 5.28 applies. According to clause (a) we need to show for all $c \in C$ and $c' \in C$ that $(c \Rightarrow c') \leqslant [\![c]\!]_{\emptyset}([c'])$. Actually, we can show immediately the stricter property that $(c \Rightarrow c') = [\![c]\!]_{\emptyset}([c'])$ in case of constants. We show this for number constants $n_i \in C_{num}$ and $n_j \in C_{num}$ only. We have

$$n_i \Rightarrow n_j \overset{(i)}{=} \begin{cases} 1\,, n_j = n_i \\ 0\,, else \end{cases} \overset{(ii)}{=} \begin{cases} 1\,, j = i \\ 0\,, else \end{cases} \overset{(iii)}{=} [\![n_i]\!]_{\rho}(j) \overset{(iv)}{=} [\![n_i]\!]_{\rho}[n_j] \tag{5.295}$$

In Eqn. (5.295), (i) holds due to Lemma 5.30, (ii) holds due to the definition of number constants C_{num} in Sect. 3.1.1, (iii) holds due to the semantic equation Eqn. (5.24) and (iv) holds due to the definition of data points in Def. 5.25.

In Case of $+1(M)$***,*** $-1(M)$ ***and*** $0?(M)$: We show this for terms of the form $+1(M)$ only. The two other cases $-1(M)$ and $0?(M)$ can be proven analogously. Given a term $+1(M)$, we can distinguish two cases, i.e., the case in which M is closed and the case in which M is an open term with some free variables $x_1, \ldots, x_n$. If the term $+1(M)$ is closed, we know that also M is closed and we can immediately exploit the induction hypothesis $\alpha(M)$. Due to $\alpha(M)$ and the fact that M is closed we can apply the performance lemma for basic operators, i.e., Lemma 5.33, so that we have that $\alpha(+1(M))$. In case the term is open we need to consider all lists of closed, performing terms $N_1, \ldots, N_n$ such that $+1(M)[x_1 := N_1] \ldots [x_n := N_n]$ is a closed term. However, in such a case, we know that also $M[x_1 := N_1] \ldots [x_n := N_n]$ is a closed term. Furthermore, due to $\alpha(N_1), \ldots, \alpha(N_n)$ we also know that $\alpha(M[x_1 := N_1] \ldots [x_n := N_n])$. Altogether, we can again apply Lemma 5.33, so that we have $\alpha(+1(M))$ also in this case.

The other cases follow analogously. We always follow the pattern that we have seen in the case of basic operators. We distinguish two cases, i.e., the case in which the constituting terms are already closed and the case in which the constituting terms are open. Actually, the case of closed constituting terms can also be treated as a special case of the case of open constituting terms with an empty list of free variables, so that the explicit treatment of the two cases can be considered a bit artificial. However, in both cases we eventually apply the corresponding helper lemma elaborated in the preceding text. In case of conditional expressions we apply Lemma 5.35, in case of applications we apply Lemma 5.36, in case of abstraction we apply Lemma 5.38, in case of recursion we apply Lemma 5.51 and, finally, in case of probabilistic choices, we apply Lemma 5.53. □

Theorem 5.55 (Semantic Correspondence) *For each program M and constant c we have the following:*

$$[\![M]\!]_{\emptyset}[c] = \big(M \Rightarrow c\big) \qquad (5.296)$$

Proof. The theorem follows immediately as a corollary from Lemma 5.27 and Lemma 5.54. □

5.4 Important Readings on Domains and Probabilism

In [242] Dana Scott clarifies the semantics of the stochastic lambda calculus. He achieves this by using random variables to model randomized oracles and adding them to his graph model [240] of the lambda calculus.

In his seminal work [232], Saheb-Djahromi defines an operational semantics and denotational semantics for the probabilistic lambda calculus; compare also with Sect. 3.4. The denotational semantics is provided based on the notion of probabilistic domains [233]. In [233], a probabilistic domain is defined by the probabilistic distributions over a domain. The probabilistic distributions are partially ordered with respect to the open sets of the underlying domain, i.e., approximation between two distributions is defined pointwise for all open sets. In [233] two alternative options are proposed as base domains for the number type of the probabilistic lambda calculus. One option is the probabilistic domain over the natural numbers $(\mathbb{N}_0, \leqslant, 0)$ with their natural ordering and 0 as bottom element as cpo. The other option is the flat domain $(\mathbb{N}_0, \sqsubseteq, \bot)$. This option is also used for the denotational semantics in [232]. The probabilistic domains over flat domains in [232] and the constructions based on pre-distributions used in this book are isomorphic. Probabilistic distributions over flat domains are sufficient for the denotational semantics in [232]. This is so, because probabilistic choice is introduced at the level of ground types only and call-by-name does not require probabilistic distributions at higher types. Also, for the concrete call-by-value abstraction construct of the language in [232] the distributions over flat domains are sufficient, because the chosen call-by-value abstraction is a limited form of abstraction that is available for variables of ground type only.

In [66], de Frutos Escrig introduces and investigates three kind of probabilistic powerdomains on the basis of the category SFP (Sequences Finite Partial) [221], which is drawn from limits of infinite sequences of finite dcpos. One of the probabilistic powerdomains is constructed as a probability distribution over a domain. The other generalizes Smyth's notion of generating trees [246] to probabilistic generating trees. The third one achieves a probabilistic version of Scott's domain-theoretic notion of information system [241]. In [67], de Frutos Escrig provides a model based on a reductionist notion of control systems.

For closure properties of the category of retracts from SFP, i.e., the category of finitely continuous partially ordered sets, and its exploitation in the semantics of probabilistic computations see Graham [113].

In [168, 169] Dexter Kozen introduces two interpretations for probabilistic programming languages, i.e., based on partial measurable functions on a measurable space on the one hand, and continuous linear operators on a partially ordered Banach space on the other hand. Also, natural embeddings of standard semantic domains into these structures are shown.

In [143, 142] Jones and Plotkin construct a probabilistic powerdomain on the basis of directed complete partial orders. The powerdomain construction forms a monadic functor on the category of the involved domains. On the basis of this, the work gives an axiomatic semantics for a probabilistic imperative programming language and a denotational semantics to a call-by-value version [142] of the probabilistic lambda calculus that is based on the computational lambda calculus [197] of Eugenio Moggi.

In [130] Heckmann defines a powerdomain construction on the basis of information systems as defined by Vickers [259]. He shows that this powerdomain construction is equivalent to the powerdomain construction defined by Jones and Plotkin in [143, 142].

In [146] Jung and Tix deal with the problem that the category of continuous domains, which is the host of the continuous powerdomain construction from [143, 142], is not a Cartesian closed category [173, 144]. Against this background, the paper provides a review of Graham's approach [113] to restrict the powerdomain construction to the retracts of bifinite domains [221]; see also [1]. Next, two solutions are approached based on finite trees on one hand, and finite reversed trees on the other hand. Also, a promising solution based on Lawson-compact domains, see [144, 105], is investigated. In [145] Jung investigates the relationship of stably compact spaces [106] and Nachbin's compact ordered spaces [206, 207] and their relevance to the construction of probabilistic powerdomains.

In [258], Tix, Keimel and Plotkin elaborate a denotational semantics for programming languages that support both non-deterministic and probabilistic choice. The probabilistic powerdomain of [143, 142] is taken as a starting point and blended with each of the three standard powerdomains used for the interpretation of non-determinism, i.e., upper, lower and convex powerdomains; see [221, 222, 246, 247]. The resulting probabilistic powerdomains are made the subject of investigation of Hahn-Banach-style theorems [147]. It is demonstrated that they work for programming languages that combine both non-deterministic and probabilistic choice, i.e., they are used for the denotational semantics of the concrete imperative programming language introduced by Morgan and McIver in [184].

In [63], Danos and Ehrhard establish a model of linear logic [108] based on a notion of probabilistic coherent spaces PCS*s*; compare also with [109]. Then, they use PCS*s* to define a denotational semantics for the probabilistic typed lambda calculus that they define by extending PCF [223] by a random generator primitive. Furthermore, in [96] Ehrhard, Pagani and Tasson use PCS*s* also to give an adequate model to a probabilistic version of the pure, i.e., untyped, lambda calculus. Again, in [98], they investigate the PCS-based denotational semantics of the probabilistic typed lambda calculus and show that it is fully abstract; see also [97] for a detailed version of [98].

In [20, 21] Barker defines a probabilistic version of PCF by adding a programming primitive for random choice. The random choice allows for arbitrary measures organized on Cantor trees. A monadic small-step semantics as well as a monadic big-step semantics against a driving input bit stream is provided. A full Haskell implementation of the proposed PCF extension is provided; see [269]. The random choice monad turns out to be an endofunctor on the Cartesian closed category of bc-domains [240]; compare also with the work of Goubault-Larrecq and Varacca in [112] as well as the analysis in [194, 195] by Michael Mislove.

In [196] Michael Mislove achieves a domain-theoretic foundation of random variables. The work complements and transcends the important strand of research on random variables and domains embodied in [194, 195]. In particular, the work achieves an adjustment of Skorohod's Theorem [252] to subprobabilities as the crucial domain-theoretic objects.

References

1. Samy Abbes and Klaus Keimel. Projective Topology on Bifinite Domains and Applications. Theoretical Computer Science, vol. 365, no. 3, 2006, pp. 171–183.
2. Samson Abramsky. The Lazy λ-Calculus. In D. Turner, ed.: Research Topics in Functional Programming, Addison-Wesley, Boston, 1990, pp. 65–117.
3. Samson Abramsky and Achim Jung. Domain Theory. In D.M. Gabbay, T.S.E. Maibaum, eds.: Handbook of Logic in Computer Science, vol. 3, Clarendon Press, Oxford, 1996.
4. Peter Aczel. An Introduction to Inductive Definitions. In Jon Barwise, ed.: Handbook of Mathematical Logic – Studies in Logic and the Foundations of Mathematics, vol. 90, Elsevier, Amsterdam, 1977, pp. 739–782.
5. Alfred V. Aho, Monica S. Lam, Ravi Sethi, and Jeffrey D. Ullman. Compilers – Principles, Techniques, and Tools, 2nd ed. Addison-Wesley, Boston, 2006.
6. Charalambos D. Aliprantis, and Owen Burkinshaw. Principles of Real Analysis, 3rd Edition, Academic Press, Cambridge, 1998.
7. Dagmar Auer, Dirk Draheim, and Verena Geist. Extending BPMN with Submit/Response-Style User Interaction Modeling. In: Proceedings of BPMN Workshop at CEC'09 – the 11th IEEE Conference on Commerce and Enterprise Computing, 2009, pp. 368–374.
8. Colin Atkinson and Dirk Draheim. Cloud-Aided Software Engineering - Evolving Viable Software Systems through a Web of Views. In Zaigham Mahmood, Saqib Saeed, eds.: Software Engineering Frameworks for the Cloud Computing Paradigm. Springer, 2013, pp. 255–281.
9. Colin Atkinson, Philipp Bostan, and Dirk Draheim. Foundational MDA Patterns for Service-Oriented Computing. In: The Journal Of Object Technology, vol. 14, no. 1, 2015, pp. 1–30.
10. Colin Atkinson, Dirk Draheim, and Verena Geist. Typed Business Process Specification. In: Proceedings of EDOC'2010 – the 14th IEEE International Enterprise Computing Conference, IEEE Computer Society, 2010, pp. 69–78
11. Adnan Aziz, Kumud Sanwal, Vigyan Singhal, and Robert K. Brayton. Verifying Continuous Time Markov Chains. In Rajeev Alur, Thomas Henzinger, eds.: Proceedings of CAV'96 – the 8th International Conference on

Computer-Aided Verification, LNCS 1102, Springer, Berlin, 1996, pp. 269–276.

12. Adnan Aziz, Kumud Sanwal, Vigyan Singhal, and Robert K. Brayton. Model Checking Continuous Time Markov Chains. ACM Transactions on Computational Logic, vol. 1, no. 1, July 2000, pp. 162–170.
13. Soheib Baarir, Marco Beccuti, Davide Cerotti, Massimiliano De Pierro, Susanna Donatelli, and Giuliana Franceschinis. The GreatSPN Tool – Recent Enhancements. ACM SIGMETRICS Performance Evaluation Review, vol. 36, no. 4, March 2009, pp. 4–9.
14. Christel Baier, Boudewijn Haverkort, Holger Hermanns, and Joost-Pieter Katoen. Model Checking Continuous-Time Markov Chains by Transient Analysis. In E. Allen Emerson, Aravinda Prasad Sistla, eds.: Proceedings of CAV 2000 – the 12th Annual Symposium on Computer-Aided Verification, LNCS 1855, Springer, 2000, pp. 358–372.
15. Christel Baier, Holger Hermanns, Joost-Pieter Katoen, and Boudewijn R. Haverkort. Efficient Computation of Time-Bounded Reachability Probabilities in Uniform Continuous-Time Markov Decision Processes. Theoretical Computer Science, vol. 345, no. 1, pp. 2–26, 2005.
16. Jørgen Bang-Jensen and Gregory Gutin. Digraphs – Theory, Algorithms and Applications, Springer, Berlin, 2009.
17. Rodrigo Bañuelos. Lecture Notes on Measury Theory and Probability. Purdue University, June 2003.
18. Henk P. Barendregt. The Lambda Calculus – Its Syntax and Semantics. North Holland, Amsterdam, 1984.
19. Henk P. Barendregt. Lambda Calculi with Types. In S. Abramsky, D.M. Gabbay, T.S.E. Maibaum, eds.: Handbook of Logic in Computer Science, vol. 2. Oxford University Press, Oxford, 1992.
20. Tyler C. Barker. A Monad for Randomized Algorithms. Ph.D. Thesis, Tulane University, April 2016.
21. Tyler C. Barker. A Monad for Randomized Algorithms. In: Proceedings of MFPS'2016 – the 32nd Conference on the Mathematical Foundations of Programming Semantics, Electronic Notes in Theoretical Computer Science, vol. 325, October 2016, pp. 47–62.
22. A.K. Basu. Measure Theory and Probability. Prentice-Hall of India, 2004.
23. Klaus J. Berkling and Elfriede Fehr. A Consistent Extension of the Lambda-Calculus as a Base for Functional Programming Languages. In: Information and Control, vol. 55, no. 1–3, 1982, pp. 89–101.
24. Klaus J. Berkling and Elfriede Fehr. A Modification of the Lambda-Calculus as a Base for Functional Programming Languages. In M. Nielsen, E.M. Schmidt, eds.: Proceedings of ICALP'82 – the 9th International Colloquium on Automata, Languages and Programming, LNCS 140, Springer, Berlin, 1982, pp. 35–47.
25. Ethan Bernstein and Umesh Vazirani. Quantum Complexity Theory. Society for Industrial and Applied Mathematics (SIAM) Journal on Computing, vol. 26, no. 5, October 1997, pp. 1411–1473.
26. Evert Willem Beth. Semantic Entailment and Formal Derivability. Mededelingen van de Koninklijke Nederlandse Akademie van Wetenschappen Afdeling Letterkunde, vol. 18, no. 3, 1955, pp.309–342.
27. Evert Willem Beth. The Foundations of Mathematics, North Holland Publishing, Amsterdam, 1959.

28. B.R. Bhat. Modern Probability Theory. New Age International Publishers, New Delhi, 2009.
29. Wayne D. Blizard. Multiset Theory. Notre Dame Journal of Formal Logic, vol. 30, no. 1, 1989, pp. 36–66.
30. Eckard Bode, Marc Herbstritt, Holger Hermanns, Sven Johr, Thomas Peikenkamp, Reza Pulungan, Ralf Wimmer, and Bernd Becker. Compositional Performability Evaluation for STATEMATE. In: Proceedings of QEST'06 – the 3rd International Conference on the Quantitative Evaluation of Systems, IEEE Computer Society, 2006, pp. 167–178.
31. Gunter Bolch, Stefan Greiner, Hermann de Meer, and Kishor S. Trivedi. Queueing Networks and Markov Chains – Modeling and Performance Evaluation with Computer Science Applications, 2nd edition. Wiley, Chichester, May 2006.
32. John Alan Bondy and U.S.R. Murty. Graph Theory with Applications. North Holland, Amsterdam, 1976.
33. John Alan Bondy and U.S.R. Murty. Graph Theory. Springer, Berlin, 2008.
34. Behzad Bordbar, Dirk Draheim, Matthias Horn, Ina Schulz, and Gerald Weber. Integrated Model-Based Software Development, Data Access and Data Migration. In Lionel Briand, Clay Williams, eds.: Model Driven Engineering Languages and Systems, LNCS 3713, Springer, Berlin, 2005, pp. 382–396.
35. Johannes Borgström, Ugo Dal Lago, Andrew D. Gordon, and Marcin Szymczak. A Lambda-Calculus Foundation for Universal Probabilistic Programming. In: Proceedings of ICFP'2016 – the 21st ACM SIGPLAN International Conference on Functional Programming, 2016, pp. 33–46.
36. Johannes Borgström, Ugo Dal Lago, Andrew D. Gordon, and Marcin Szymczak. A Lambda-Calculus Foundation for Universal Probabilistic Programming – v.4. In: The Computing Research Repository, arXiv:1512.08990, Cornell University Library, May 2016, pp. 1–56.
37. Leo Breiman. Probability. Addison-Wesley, Boston, 1968, out of print, appeared also as: Probability – Classics in Applied Mathematics, vol. 7, Society for Industrial and Applied Mathematics, Philadelphia, 1992.
38. Franck van Breugel, Michael Mislove, Joël Ouaknine, and James Worrell. Domain Theory, Testing and Simulation for Labelled Markov Processes. Theoretical Computer Science, vol. 333, no. 1–2, March 2005, pp. 171–197.
39. Bruno Buchberger. Mathematics of 21st Century – A Personal View. In L. Kovacs, T. Kutsia, eds.: Proceedings of SCSS 2013 – the 4th Symposium on Symbolic Computation in Software Science, EPiC Series, vol. 15, 2013, pp. 1–1.
40. Bruno Buchberger et.al. Theorema – Towards Computer-Aided Mathematical Theory Exploration. In: Journal of Applied Logic, vol. 4, no. 14, 2005, pp. 470-504.
41. Juanito Camilleri and Tom Melham. Reasoning with Inductively Defined Relations in the HOL Theorem Prover. Technical Report No. 265, University of Cambridge Computer Laboratory, August 1992.
42. Luca Cardelli. Type Systems. In Allan B. Tucker, ed.: The Computer Science and Engineering Handbook. CRC Press, Boca Raton, 1997.
43. James B. Carrell. Fundamentals of Linear Algebra, July 2005. `https://www.math.ubc.ca/~carrell/NB.pdf`

44. Christos G. Cassandras, Michael I. Clune, and Pieter J. Mosterman. Hybrid System Simulation with SimEvents. In: Proceedings of the 2nd IFAC Conference on Analysis and Design of Hybrid Systems, IFAC Proceedings Volumes, vol. 39, no. 5, Elsevier, Amsterdam, 2006, pp. 267–269.
45. Kai Lai Chung. Elementary Probability Theory with Stochastic Processes. Springer, Berlin, 1974.
46. Kai Lai Chung. A Course in Probability Theory, revised 2nd edition. Academic Press, Cambridge, 2001.
47. Alonzo Church. A Set of Postulates for the Foundation of Logic. Annals of Mathematics, Series 2, no. 33, 1932, pp. 346–366.
48. Alonzo Church. An Unsolvable Problem of Elementary Number Theory. American Journal of Mathematics, vol. 58, no. 2, April 1936, pp. 345–363.
49. Alonzo Church, John B. Rosser. Some Properties of Conversion. Transactions of the American Mathematical Society, vol. 39, no. 3, 1936, pp. 472–483.
50. Allan Clark, Stephen Gilmore, Jane Hillston, and Mirco Tribastone. Stochastic Process Algebras. In Marco Bernardo, Jane Hillstone, eds.: Formal Methods for Performance Evaluation, LNCS 4486, Springer, 2007, pp. 132–179.
51. Thierry Coquand and Peter Dybjer. Inductive Definitions and Type Theory – an Introduction. In P.S. Thiagarajan, ed.: Foundation of Software Technology and Theoretical Computer Science, Lecture Notes in Computer Science 880, Springer, Berlin, 1994, pp. 60–76.
52. Thomas H. Cormen, Charles E. Leiserson, Ronald L. Rivest, and Clifford Stein. Introduction to Algorithms, 3rd edition. MIT Press, Cambridge, 2009.
53. Patrick Cousot and Michael Monerau. Probabilistic Abstract Interpretation. In Helmut Seidl, ed.: Proceedings of ESOP'12 – the 21st European Symposium on Programming, LNCS 7211, Springer, 2012, pp. 169–193.
54. Raphaëlle Crubillé and Ugo Dal Lago. On Probabilistic Applicative Bisimulation and Call-by-Value λ-Calculi. In Zhang Shao, ed.: Proceedings of ESOP – The 23rd European Symposium on Programming, LNCS 8410, Springer, Berlin, 2014, pp. 209–228.
55. Raphaëlle Crubillé and Ugo Dal Lago. On Probabilistic Applicative Bisimulation and Call-by-Value λ-Calculi. In: The Computing Research Repository, arXiv:1401.3766v2, Cornell University Library, January 2104, pp. 1–30.
56. Ugo Dal Lago and Paolo Parisen Toldin. A Higher-Order Characterization of Probabilistic Polynomial Time. In R. Peña, M. van Eekelen, O. Shkaravska, eds.: Proceedings of FOPARA 2011 – the 2nd International Workshop on Foundational and Practical Aspects of Resource Analysis. LNCS 7177, Springer, Berlin, 2012, pp. 1–18.
57. Ugo Dal Lago and Paolo Parisen Toldin. A Higher-Order Characterization of Probabilistic Polynomial Time. Information and Computation, vol. 241, April 2015, pp. 114–141.
58. Ugo Dal Lago and Margherita Zorzi. Probabilistic Operational Semantics for the Lambda Calculus. RAIRO – Theoretical Informatics and Applications, vol. 46, no. 3, 2012, pp. 413–450.

59. Ugo Dal Lago and Sara Zuppiroli. Probabilistic Recursion Theory and Implicit Computational Complexity. In G. Ciobanu, D. Méry, eds.: Proceedings of ICTAC 2014 – 11th International Colloquium on Theoretical Aspects of Computing, LNCS 8687, Springer, Berlin, September 2014, pp. 97–114
60. Ugo Dal Lago and Sara Zuppiroli. Probabilistic Recursion Theory and Implicit Computational Complexity. In: The Computing Research Repository, arXiv:1406.3378, Cornell University Library, June 2104, pp. 1–27.
61. Ugo Dal Lago, Davide Sangiorgi, and Michele Alberti. On Coinductive Equivalences for Higher-Order Probabilistic Functional Programs. ACM SIGPLAN Notices – Proceedings of POPL'14 – the 41st ACM SIGPLAN-SIGACT Symposium on Principles of Programming Languages, vol. 49, no. 1, January 2014, pp. 297–308.
62. Ugo Dal Lago, Davide Sangiorgi, and Michele Alberti. On Coinductive Equivalences for Higher-Order Probabilistic Functional Programs. In: The Computing Research Repository, arXiv:1311.1722, Cornell University Library, November 2013, pp. 1–47.
63. Vincent Danos and Thomas Ehrhard. Probabilistic Coherence Spaces as a Model of Higher-Order Probabilistic Computation. Information and Computation, vol. 209, no. 6, June 2011, pp. 966-991.
64. Davide D'Aprile, Susanna Donatelli, and Jeremy Sproston. CSL Model Checking for the GreatSPN Tool. In C. Aykanat, T. Dayar, I. Korpeoglu, eds.: Proceedings of ISCIS 2004 – the 19th International Symposium Computer and Information Sciences, LNCS 3280, Springer, 2004, pp. 543–552.
65. Nicolaas G. de Bruijn. A Survey of the Automath Project. In J.R. Hindley, J.P. Seldin, eds.: To H.B. Curry – Essays on Combinatory Logic, Lambda-Calculus and Formalism. Academic Press, Cambridge, 1980, pp. 580–606.
66. David de Frutos Escrig. Some Probabilistic Powerdomains in the Category SFP. In Burkhard Monien, Guy Vidal-Nagnet, eds.: Proceedings of STACS'86 – the 3rd Annual Symposium on Theoretical Aspects of Computer Science. LNCS 210, Springer, Berlin, 1986, pp. 49–59.
67. David de Frutos Escrig. Probabilistic Ianov's Schemes. Theoretical Computer Science, vol. 53, no. 1, 1987, pp. 67–97.
68. Karel De Leeuw, Edward F. Moore, Claude E. Shannon, and Norman Shapiro. Computability by Probabilistic Machines. In: Automata Studies, vol. 34, 1955, pp. 183–212.
69. W.P. DeRoever. Call-by-Value versus Call-by-Name – a Proof-Theoretic Comparison. In A. Blikle, ed.: Mathematical Foundations of Computer Science, Lecture Notes in Computer Science 28, Springer, Berlin, 1975, pp. 451–463.
70. Reinhard Diestel. Graph Theory. Springer, Berlin, 2005.
71. Alessandra Di Pierro, Chris Hankin, and Herbert Wiklicky. Probabilistic λ-Calculus and Quantitative Program Analysis. Journal of Logical Computation, vol. 15, no. 2, 2005, pp. 159–179.
72. Alessandra Di Pierro, Chris Hankin, and Herbert Wiklicky. Probabilistic Semantics and Program Analysis. In A. Aldini, M. Bernardo, A. Di Pierro, H. Wiklicky, eds.: Formal Methods for Quantitative Aspects of Programming Languages, Lecture Notes in Computer Science 6154, Springer, Berlin, 2010, pp. 1–42.

73. Dirk Draheim. SEKE 2007 Invited Talk: Towards Seamless Workflow and Dialogue Specification. In: Proceedings of SEKE 2007 – The 19th International Conference on Software Engineering and Knowledge Engineering, 2007, pp. 402–403.
74. Dirk Draheim. Business Process Technology - A Unified View on Business Processes, Workflows and Enterprise Applications. Springer, Berlin, September 2010.
75. Dirk Draheim. The Service-Oriented Metaphor Deciphered. Journal of Computing Science and Engineering, vol. 4, no. 4, 2010, pp. 253–275
76. Dirk Draheim. Smart Business Process Management. In: 2011 BPM and Workflow Handbook, Digital Edition. Future Strategies, Workflow Management Coalition, 2012, pp. 207–223.
77. Dirk Draheim. MDHPCL 2012 Invited Talk. CASE 2.0 - On Key Success Factors for Cloud-Aided Software Engineering. In: Proceedings of MDHPCL – the 1st International Workshop on Model-Driven Engineering for High Performance and Cloud Computing, ACM Press, New York, 2012, pp. 1–6.
78. Dirk Draheim. Sustainable Constraint Writing and Symbolic Viewpoints of Modeling Languages. Invited Talk. In H. Decker et al., eds.: Proceedings of DEXA'14 – the 25th International Conference on Database and Expert Systems Applications, LNCS 8644, Springer, Berlin, 2014, pp. 12-19.
79. Dirk Draheim. Reflective Constraint Writing. In A. Hameurlain et al., eds.: Transactions on Large-Scale Data- and Knowledge-Centered Systems, vol. 24, LNCS 9510, Springer, Berlin, 2016, pp. 1–60.
80. Dirk Draheim. FDSE Invited Talk: The Present and Future of Large-Scale Systems Modeling and Engineering. In: Proceedings of FDSE 2106 – the 3rd International Conference on Future Data and Security Engineering. LNCS 10018, Springer, Berlin, 2016.
81. Dirk Draheim and Gerald Weber. Strongly Typed Server Pages. In Alon Halevy, Avigdor Gal, eds.: Proceedings of NGITS 2002 – Next Generation Information Technologies and Systems, LNCS 2382, Springer, Berlin, 2002, pp. 29–44.
82. Dirk Draheim and Gerald Weber. Storyboarding Form-Based Interfaces. In: Proceedings of INTERACT 2003 – the 9th IFIP TC13 International Conference on Human-Computer Interaction. IOS Press, Amsterdam, 2003, pp. 343–350.
83. Dirk Draheim and Gerald Weber. Form-Oriented Analysis – A New Methodology to Model Form-Based Applications. Springer, Berlin, October 2004.
84. Dirk Draheim and Gerald Weber (Editors). Post-Proceedings of TEAA 2006 – the VLDB Workshop on Trends in Enterprise Application Architecture, LNCS 3888, Springer, Berlin, March 2006.
85. Dirk Draheim and Gerald Weber (Editors). Post-Proceedings of TEAA 2007 – the 2nd International Conference on Trends in Enterprise Application Architecture, LNCS 4473, Springer, Berlin, June 2007.
86. Dirk Draheim and Christine Nathschläger. A Context-Oriented Synchronization Approach. In: Electronic Proceedings of PersDB 2008 – the 2nd International Workshop in Personalized Access, Profile Management, and Context Awareness: Databases (in Conjunction with the 34th VLDB Conference), 2008, pp. 20–27.

87. Dirk Draheim, Elfriede Fehr, and Gerald Weber. JSPick – a Server Pages Design Recovery Tool. In: Proceedings of CSMR 2003 – the 7th European Conference on Software Maintenance and Reengineering. IEEE Press, Los Alamitos, 2003.
88. Dirk Draheim, Melanie Himsl, Daniel Jabornig, Josef Küng, Werner Leithner, Peter Regner, and Thomas Wiesinger. Concept and Pragmatics of an Intuitive Visualization-Oriented Metamodeling Tool. Journal of Visual Languages and Computing, vol. 21, no. 4, 2010, pp. 157–170.
89. Dirk Draheim, Matthias Horn, and Ina Schulz. The Schema Evolution and Data Migration Framework of the Environmental Mass Database IMIS. In: Proceedings of SSDBM 2004 – the 16th International Conference on Scientific and Statistical Database Management. IEEE Computer Society, Los Alamitos, 2004, pp. 341–344
90. Dirk Draheim, Christof Lutteroth, and Gerald Weber. A Type System for Reflective Program Generators. In Robert Glück, Michael Lawry, eds.: Proceedings of GPCE 2005 – Generative Programming and Component Engineering, LNCS 3676, Springer, 2005.
91. Dirk Draheim, Christof Lutteroth, and Gerald Weber. Generative Programming for C#. ACM SIGPLAN Notices, vol. 40, no. 8., August 2005, pp. 29–33.
92. Dirk Draheim, Christof Lutteroth, and Gerald Weber. Integrating Code Generators into the C# Language. In: Proceedings of ICITA 2005 – the 3rd International Conference on Information Technology and Applications. IEEE Computer Society, Los Alamitos, 2005, pp. 107–110.
93. R.M. Dudley. Real Analysis and Probability. Cambridge University Press, Cambridge, 2004.
94. Marie Duflot, Laurent Fribourg, and Claudine Picaronny. Randomized Finite-State Distributed Algorithms as Markov Chains. In Jennifer Welch, ed.: Proceedings of DISC'01– the 15th International Conference on Distributed Computing, LNCS 2180, Springer, Berlin, 2008, pp. 240–254.
95. Rick Durrett. Probability – Theory and Examples. Cambridge University Press, Cambridge, 2010.
96. Thomas Ehrhard, Michele Pagani, and Christine Tasson. The Computational Meaning of Probabilistic Coherent Spaces. In: Proceedings of LICS 2011 – the 26th Annual IEEE Symposium on Logic in Computer Science, IEEE Computer Society, Los Alamitos, 2011, pp. 87–96.
97. Thomas Ehrhard, Michele Pagani, and Christine Tasson. Full Abstraction for Probabilistic PCF. In: The Computing Research Repository, arXiv:1511.01272, Cornell University Library, November 2015.
98. Thomas Ehrhard, Christine Tasson, and Michele Pagani. Probabilistic Coherence Spaces are Fully Abstract for Probabilistic PCF. In: Proceedings of POPL'14 – the 41st Annual ACM SIGPLAN-SIGACT Symposium on Principles of Programming Languages, ACM Press, New York, January 2014, pp. 309–320.
99. Hartmut Ehrig and Bernd Mahr. Fundamentals of Algebraic Specifications 1 – Equations and Initial Semantics. Springer, Berlin, 1985.
100. European Commission. Unleashing the Potential of the Cloud in Europe. Communication from the Commission to the European Parliament, the Council, the European Economic and Social Committee and the Committee

of the Regions, Com(2012) 529 final. European Commission, Brussels, 27 Sept 2012.
101. Agner K. Erlang. The Theory of Probabilities and Telephone Conversations. Nyt Tidsskrift for Matematik B, vol 20, 1909.
102. Elfriede Fehr. Expressive Power of Typed and Type-Free Programming Languages. In: Theoretical Computer Science, vol. 33, nos. 2–3, 1984, pp. 195–238.
103. Elfriede Fehr. Semantik von Programmiersprachen. Springer, Berlin, 1989.
104. Erich Gamma et al. Design Patterns – Elements of Reusable Object-Oriented Software. Addison-Wesley, Boston, 1995.
105. Gerhard Gierz, Karl Heinrich Hofmann, Klaus Keimel, Jimmie D. Lawson, Michael Mislove, and Dana S. Scott. A Compendium of Continuous Lattices. Springer, Berlin, 1980.
106. Gerhard Gierz, Karl Heinrich Hofmann, Klaus Keimel, Jimmie D. Lawson, Michael Mislove, and Dana S. Scott. Continuous Lattices and Domains. Encyclopedia of Mathematics and its Applications, vol. 93. Cambridge University Press, Cambridge, 2003.
107. Stephen Gilmore and Jane Hillston. The PEPA Workbench – a Tool to Support a Process Algebra-based Approach to Performance Modelling. In Günther Haring, Gabriele Kotsis, eds.: In Proceedings of the 7th International Conference on Computer Performance Evaluation Modelling Techniques and Tools, LNCS 794, Springer, Berlin, 1994, pp. 353–368.
108. Jean-Yves Girard. Linear Logic. Theoretical Computer Science, vol. 50, no. 1, 1987 pp. 1–101.
109. Jean-Yves Girard. Between Logic and Quantic – a Tract. In T. Ehrhard, J.-Y. Girard, P. Ruet, P. Scott, eds.: Linear Logic in Computer Science, London Mathematical Society Lecture Note Series, vol. 316, Cambridge University Press, Cambridge, 2004, pp. 346–381.
110. Noah Goodman, Vikash Mansinghka, Daniel M. Roy, Keith Bonawitz, and Joshua B. Tenenbaum. Church – a Language for Generative Models. In D.A. McAllester, P. Myllymäki, eds.: Proceedings of UAI 2008 – the 24th Conference in Uncertainty in Artificial Intelligence, AUAI Press, 2008, pp. 220–229.
111. Michael J. Gordon, Arthur J.R.G. Milner, and Christopher P. Wadsworth. Edinburgh LCF – A Mechanized Logic of Computation. Lecture Notes in Computer Science 78, Springer, Berlin, 1979.
112. Jean Goubault-Larrecq and Daniele Varacca. Continuous Random Variables. In: Proceedings of LICS'11 – the 26th Annual IEEE Symposium on Logic in Computer Science, IEEE Computer Society Press, Los Alamitos, June 2011, pp. 97–106.
113. Steven K. Graham. Closure Properties of a Probabilistic Domain Construction. In M. Main, A. Melton, M. Mislove, D. Schmidt, eds.: Proceedings of MFPS'87 – the 3rd Workshop on Mathematical Foundations of Programming Language Semantics, LNCS 298, Springer, Berlin, 1988, pp. 213–233
114. Michael A. Gray. Discrete Event Simulation – A Review of SimEvents. Computing Science Engineering, vol. 9, no. 6, 2007, pp. 62–66.
115. Donald Gross, John F. Shortle, James M. Thompson, and Carl M. Harris. Fundamentals of Queueing Theory. John Wiley & Sons, New York, 2008. s

116. Carl A. Gunter and Dana S. Scott. Semantic Domains. In Jan van Leeuwen, ed.: Handbook of Theoretical Computer Science, vol. B – Formal Models and Sematics. MIT Press, Cambridge, 1990, pp. 634–674.
117. Carl A. Gunter, Peter D. Mosses, and Dana S. Scott. Semantic Domains and Denotational Semantics. Lecture Notes of the International Summer School on Logic, Algebra and Computation. Marktoberdorf, June/August 1989. Appeared also in Jan van Leeuwen, ed.: Handbook of Theoretical Computer Science, vol. B – Formal Models and Sematics. MIT Press, Cambridge, 1990, pp. 575–631, 634–674.
118. Carl A. Gunter. Semantics of Programming Languages – Structures and Techniques. The MIT Press, Cambridge, 1992.
119. Peter J. Haas. Stochastic Petri Nets – Modelling, Stability, Simulation. Springer, Berlin, 2002.
120. B. Hahn and C. Ballinger. Tpump in Continuous Environment – Assembling the Teradata Active Data Warehouse Series. Active Data Warehouse Center of Expertise, April 2001.
121. Hans Hansson and Bengt Jonsson. A Logic for Reasoning About Time and Reliability. Formal Aspects of Computing, vol. 6, no. 5, September 1994, pp. 512–535.
122. David Harel. Statecharts – a Visual Formalism for Complex Systems. Science of Computer Programming, vol. 8, no. 3, 1987, pp.231–274.
123. David Harel and Amon Naamad. The Statemate Semantics of Statecharts. ACM Transactions on Software Engineering and Methodology, vol. 5, no. 4, 1996, pp. 293–333.
124. John Harrison. Inductive Definitions – Automation and Application. In Phillip J. Windley, Thomas Schubert and Jim Alves-Foss, eds.: Proceedings of the 1995 International Workshop on Higher Order Logic Theorem Proving and its Applications, LNCS 971, Springer, 1995, pp. 200–213.
125. M.J. Harry. The Vision of Six Sigma, 8 volumes. Tri Star Publishing, 1998.
126. M.J. Harry. Six Sigma: A Breakthrough Strategy for Profitability. In: Quality Progress, vol. 31, no. 5, May 1998, pp. 60–64.
127. Sergiu Hart, Micha Sharir and Amir Pnueli. Termination of Probabilistic Concurrent Programs. In: Proceedings of POPL'82 – the 9th ACM SIGPLAN-SIGACT Symposium on Principles of Programming Languages, ACM Press, New York, 1982, pp. 1–6.
128. Sergiu Hart, Micha Sharir and Amir Pnueli. Termination of Probabilistic Concurrent Programs. ACM Transactions on Programming Languages and Systems, vol. 5, no. 3, 1983, pp. 356–380.
129. Felix Hausdorff. Grundzüge der Mengenlehre, Veit & Comp, Berlin, 1927.
130. Reinhold Heckmann. Probabilistic Power Domains, Information Systems, and Locales. In S. Brookes, M. Main, A. Melton, M. Mislove, D. Schmidt, eds.: Proceedings of MFPS'93 – the 9th International Conference on Mathematical Foundations of Programming Semantics, LNCS 802, Springer, Berlin, 1994, pp. 410–437
131. Holger Hermanns, Joost-Pieter Katoen, Joachim Meyer-Kayser, and Markus Siegle. A Markov Chain Model Checker. In Susanne Graf, Michael Schwartzbach, eds.: Proceedings of TACAS 2000 – the 6th International Conference on Tools and Algorithms for the Construction and Analysis of Systems, LNCS 1785, Springer, Berlin, 2000, pp. 347–362.

132. Carl Hewitt and Henry Baker. Actors and Continuous Functionals. Technical Report MIT/LCS/TR-194. Laboratory of Computer Science, Massachusetts Institute of Technology, 1977.
133. Carl Hewitt, Peter Bishop, and Richard Steiger. A Universal Modular ACTOR Formalism for Artificial Intelligence. In: Proceedings of IJCAI'73 – the 3rd International Joint Conference on Artificial intelligence, Morgan Kaufmann, San Francisco, 1973, pp. 235–245.
134. Jane Hillston. A Compositional Approach to Performance Modelling. Cambridge University Press, Cambridge, 1996.
135. Melanie Himsl, Daniel Jabornig, Werner Leithner, Dirk Draheim, Peter Regner, Thomas Wiesinger, and Josef Küng. An Iterative Process for Adaptive Meta- and Instance Modeling. In Roland Wagner, Norman Revell, Günter Pernul, eds.: Database and Expert Systems Applications. LNCS 4653, Springer, Berlin, 2007, pp. 519–528.
136. C.A.R. Hoare. An Axiomatic Basis for Computer Programming. Communications of the ACM, vol. 12, no. 10, 1969, pp. 576–580.
137. J.E. Hopcroft, R. Motwani, and J.D. Ullman. Introduction to Automata Theory, Languages, and Computation. Addison-Wesley, Boston, 2001.
138. P. Hudak and P. Wadler (Editors). Report on the Programming Language Haskell Version 1.1. Computer Science Departments, Glasgow University and Yale University, August 1991.
139. Joe Hurd. A Formal Approach to Probabilistic Termination. In: V.A. Carreño, C. Muñoz, S. Tahar, eds.: Proceedings of TPHOLs'2002 – the 15th International Conference on Theorem Proving in Higher Order Logics, LNCS 2410, Springer, Berlin, 2002, pp. 230–245.
140. David N. Jansen, Joost-Pieter Katoen, Marcel Oldenkamp, Mariëlle Stoelinga, and Ivan Zapreev. How Fast and Fat Is Your Probabilistic Model Checker – an Experimental Performance Comparison. In Karin Yorav, ed.: Proceedings of HVC'2007 – the 3rd International Haifa Verification Conference (Hardware and Software – Verification and Testing Volume), LNCS 4899, Springer, Berlin, 2008, pp. 69–85.
141. Edwin T. Jaynes. Probability Theory. Cambridge University Press, Cambridge, 2003.
142. Claire Jones. Probabilistic Non-Determinism. Ph.D. Thesis, University of Edinburgh, August 1989.
143. Claire Jones and Gordon D. Plotkin. A Probabilistic Powerdomain of Evaluations. In: Proceedings of the 4th Annual Symposium on Logic in Computer Science, IEEE Computer Society, Los Alamitos, 1989, pp. 186–195.
144. Achim Jung. Cartesian Closed Categories of Domains, CWI Tracts, vol. 66, Centrum voor Wiskunde en Informatica, 1989.
145. Achim Jung. Stably Compact Spaces and the Probabilistic Powerspace Construction. Electronic Notes in Theoretical Computer Science, vol. 73, 2004, pp. 1–2.
146. Achim Jung and Regina Tix. The Troublesome Probabilistic Powerdomain. Electronic Notes in Theoretical Computer Science, vol. 13, 1998, pp. 70–91.
147. Leonid V. Kantorovič, Gleb P. Akilov. Functional Analysis. Pergamon, Oxford, 1982.
148. Richard M. Karp. An Introduction to Randomized Algorithms. Journal of Discrete Applied Mathematics – Special Volume: Combinatorics and Theoretical Computer Science, vol. 34, nos. 1–3, Nov. 1991, pp. 165–201.

149. Joost-Pieter Katoen, Maneesh Khattri, and Ivan S. Zapreev. A Markov Reward Model Checker. In: Proceedings of QEST'05 – the 2nd International Conference on the Quantitative Evaluation of Systems, IEEE Computer Society, Los Alamitos, 2005, pp. 243–244.
150. Klaus Keimel, A. Rosenbusch, and Thomas Streicher. Relating Direct and Predicate Transformer Partial Correctness Semantics for an Imperative Probabilistic-Nondeterministic Language. Theoretical Computer Science, vol. 412, no. 25, 2011, pp. 2701-2713.
151. Stephen Cole Kleene. General Recursive Functions of Natural Numbers. Mathematische Annalen, vol. 112, no. 1, 1936, pp. 727–742.
152. Stephen Cole Kleene. On Notation for Ordinal Numbers. Journal of Symbolic Logic, vol. 3, no. 4, 1938, pp. 150–155
153. Stephen Cole Kleene. Recursive Functionals and Quantifiers of Finite Types (i). Transactions of the American Mathematical Society, vol. 91, 1959, pp. 1–52.
154. Stephen Cole Kleene. Recursive Functionals and Quantifiers of Finite Types (ii). Transactions of the American Mathematical Society, vol. 108, 1959, pp. 106–142.
155. Stephen Cole Kleene. Recursive Functionals and Quantifiers of Finite Types Revisited (i). In J.E. Fenstad, R.O. Gandy, E. Sacks, eds.: Generalized Recursion Theory (ii). North Holland, Amsterdam, 1978, pp. 185–222.
156. Stephen Cole Kleene. Recursive Functionals and Quantifiers of Finite Types Revisited (ii). In J. Barwise, H.J. Keisler, K. Kunen, eds.: The Kleene Symposium. North Holland, Amsterdam, 1980, pp. 1–29.
157. Stephen Cole Kleene. Recursive Functionals and Quantifiers of Finite Types Revisited (iii). In G. Metakides, ed.: The Patras Logic Symposium. North Holland, Amsterdam, 1982, pp. 1–40.
158. Stephen Cole Kleene. Unimonotone Functions of Finite Types – Recursive Functionals and Quantifiers of Finite Types Revisited (iv). In A. Nerode, R.A. Shore, eds.: Recursion Theory, AMS Proceedings of Symposia in Pure Mathematics, 1985, vol. 42, pp. 119–138.
159. Stephen Cole Kleene. Recursive Functionals and Quantifiers of Finite Types Revisited (v). In G. Metakides, ed.: The Patras Logic Symposium. Transactions of the American Mathematical Society, vol. 325, no. 2, 1991, pp. 593–630.
160. Bronisław Knaster. Un théorème sur les fonctions d'ensembles. In: Annales de la Société Polonaise Mathématiques, tome (vi) – Comptes rendu des séances de la Société Polonaise Mathématiques Section de Varsovie. Roczniki Polskiego Towarzystwa Matematycznego, 1927, pp. 133–134.
161. Daphne Koller, David McAllester, and Avi Pfeffer. Effective Bayesian Inference for Stochastic Programs. In: Proceedings of AAAI'97 – the 14th National Conference on Artificial Intelligence, MIT Press, Cambridge, 1997.
162. Andrey Kolmogorov. Grundbegriffe der Wahrscheinlichkeitsrechnung. Springer, Berlin, 1933.
163. Andrey Kolmogorov. Foundations of the Theory of Probability. Chelsea, New York, 1956.
164. Andrey Kolmogorov. On Logical Foundations of Probability Theory. In Kiyosi Itô and Jurii V. Prokhorov, eds.: Probability Theory and Mathematical Statistics. Lecture Notes in Mathematics 1021, Springer, 1982, pp. 1–5.

165. Denes König. Sur les correspondances multivoques des ensembles. Fundamenta Mathematicae, vol. 8, 1926, pp. 114–134.
166. Denes König. Über eine Schlussweise aus dem Endlichen ins Unendliche. Acta Scientiarum Mathematicarum (Acta Szeged), vol. 3, nos. 2–3, 1927, pp. 121–130.
167. Felix Kossak, Christa Illibauer, Verean Geist, Christine Natschläger, Thomas Ziebermayr, Bernhard Freudenthaler, Theodorich Kopetzky, Klaus-Dieter Schewe. Hagenberg Business Process Modelling Method. Springer, Berlin, 2016.
168. Dexter Kozen. Semantics of Probabilistic Programs. In: Proceedings of FOCS'79 – the 20th Annual Symposium on Foundations of Computer Science, IEEE Computer Society, Los Alamitos, 1979, pp. 101–114.
169. Dexter Kozen. Semantics of Probabilistic Programs. Journal of Computer and System Sciences, vol. 22, no. 3, 1981, pp. 328–350.
170. Dexter Kozen and Alexandra Silver. Practical Coinduction. In: Mathematical Structures in Computer Science, vol. 1, Cambridge University Press, February 2016.
171. Kenneth Kunen. Set Theory – An Introduction to Independence Proofs. North Holland, Amsterdam, 1980.
172. Marta Z. Kwiatkowska, Gethin Norman and David Parker. Prism 2.0 – A Tool for Probabilistic Model Checking. In QEST'2004 – Proceedings of the 1st International Conference on the Quantitative Evaluation of Systems, IEEE Computer Society, Los Alamitos, 2004, pp. 322–323.
173. Joachim Lambek. From Lambda Calculus to Cartesian Closed Categories. In J. P. Seldin and J. Hindley, eds.: To H.B. Curry – Essays on Combinatory Logic, Lambda Calculus and Formalism, Academic Press, Cambridge, 1980, pp. 376–402.
174. Kim G. Larsen and Arne Skou. Bisimulation through Probabilistic Testing. Information and Computation, vol. 94, no. 1, September 1991.
175. Axel Legay, Benoît Delahaye, and Saddek Bensalem. Statistical Model Checking – an Overview. In Howard Barringer et al., eds.: Proceedings of RV 2010 – the 1st International Conference on Runtime Verification, LNCS 6418, Springer, Berlin, 2010, pp. 122–135.
176. Daniel J. Lehmann, Amir Pnueli, and Jonathan Stavi. Impartiality, Justice and Fairness – the Ethics of Concurrent Termination. In S. Even, O. Kariv, eds.: Proceedings of ICALP'81 – the 8th International Colloquium on Automata, Languages and Programming, LNCS 115, Springer, Berlin, 1981, pp. 264–277.
177. Olivier Lévêque. Lecture Notes on Markov Chains. National University of Ireland, 2011.
178. Christof Lutteroth, Dirk Draheim, and Gerald Weber. A Type System for Reflective Program Generators. Science of Computer Programming, vol. 76, no. 5, May 2011, pp. 392–422.
179. Christian Maes. An Introduction to the Theory of Markov Processes – Mostly for Physics Students. Instituut voor Theoretische Fysica, K.U. Leuven, 2012.
180. Vikash Mansinghka, Daniel Selsam, and Yura Perov. Venture – a Higher-Order Probabilistic Programming Platform with Programmable Inference. In: The Computing Research Repository, arXiv:1601.04943, Cornell University Library, April 2014, pp. 1–10.

181. Andrei Andrejewitsch Markov. Extension of the Law of Large Numbers to Dependent Quantities (in Russian). Izvestiya Fiziko-Matematicheskikh Obschestva Kazan University, vol. 15, 1906, pp. 135–156.
182. Per Martin-Löf. Constructive Mathematics and Computer Programming. In L.J. Cohen et al., eds.: Logic, Methodology and Philosophy of Science, VI, 1979, pp.153–175. North Holland, Amsterdam, 1982.
183. Per Martin-Löf. Intuitionistic Type-Theory. Bibliopolis, 1984.
184. Annabelle McIver and Carroll Morgan. Partial Correctness for Probabilistic Demonic Programs. Theoretical Computer Science, vol. 266, nos. 1–2, 2001, pp. 513–541, 2001.
185. Annabelle McIver and Carroll Morgan. Abstraction, Refinement and Proof for Probabilistic Systems. Springer, Berlin, 2005.
186. Edward J. McShane. Partial Orderings and Moore-Smith Limits. The American Mathematical Monthly, vol. 59, 1952, pp. 1–11.
187. Tom Melham. A Package for Inductive Relation Definitions in HOL. In M. Archer, J. J. Joyce, K. N. Levitt, and P. J. Windley, eds.: Proceedings of the 1991 International Workshop on the HOL Theorem Proving System and its Applications, IEEE Computer Society Press, Los Alamitos, 1992, pp. 350–357.
188. P. Mell and T. Grance. The NIST Definition of Cloud Computing – version 15. National Institute of Standards and Technology, Information Technology Laboratory, 2009.
189. Carl D. Meyer. Matrix Analysis and Applied Linear Algebra. Society for Industrial and Applied Mathematics, Philadelphia, February 2001.
190. Robin Milner. Logic for Computable Functions – Description of a Machine Implementation. Technical Report AD-785 072, Advanced Research Projects Agency National Aeronautics and Space Administration, May 1972.
191. Robin Milner. Implementation and Applications of Scott's Logic for Computable Functions. ACM SIGPLAN Notices – Proceedings of ACM Conference on Proving Assertions about Programs, vol. 7, no 1, January 1972, pp. 1–6.
192. Robin Milner. Models of LCF. Memo AIM-186, Stanford Artificial Intelligence Laboratory, Stanford University, 1973.
193. Robin Milner, Mads Tofte, Robert Harper. The Definition of Standard ML. MIT Press, Cambridge, 1990.
194. Michael Mislove. Anatomy of a Domain of Continuous Random Variables (i). In: Models of Interaction – Essays in Honour of Glynn Winskel. Theoretical Computer Science, vol. 546, August 2014, pp. 176–187.
195. Michael Mislove. Anatomy of a Domain of Continuous Random Variables (ii). In Bob Coecke, Luke Ong, Prakesh Panangaden, eds.: Computation, Logic, Games, and Quantum Foundations – the Many Facets of Samson Abramsky, LNCS 7860, Springer, Berlin, 2013, pp. 225–245.
196. Michael Mislove. Domain Theory and Random Variables. In: The Computing Research Repository, arXiv:1607.07698, Cornell University Library, August 2106, pp. 1–23.
197. Eugenio Moggi. Computational Lambda-Calculus and Monads. In: Proceedings of LICS'89 – the 4th Annual Symposium on Logic in Computer Science, IEEE Computer Society, Los Alamitos, 1989, pp. 14–23.

198. David Monniaux. Abstract Interpretation of Probabilistic Semantics. In Jens Palsberg et al., eds.: Proceedings of SAS' 2000 – the 7th International Symposium on Static Analysis, LNCS 1824, Springer, Berlin, 2000, pp. 322–339.
199. David Monniaux. Abstract Interpretation of Programs as Markov Decision Processes. Science of Computer Programming, vol. 58, no. 1–2, 2005, pp. 179–2052.
200. Eliakim H. Moore and H. L. Smith. A General Theory of Limits. American Journal of Mathematics, vol. 44, no. 2, April 1922, pp. 102–121.
201. Yiannis N. Moschovakis. Elementary Induction on Abstract Structures. Dover Publications, Mineola, 1974.
202. Yiannis N. Moschovakis. On the Basic Notions in the Theory of Induction. In Robert E. Butts and Jaako Hintikka, eds.: Logic, Foundations of Mathematics, and Computability Theory, The University of Western Ontario Series in Philosophy of Science Volume 9, D. Reidel, Dordrecht, 1977, pp. 207–236.
203. Peter D. Mosses. Denotational Semantics. In Jan van Leeuwen, ed.: Handbook of Theoretical Computer Science, vol. B – Formal Models and Sematics. MIT Press, Cambridge, 1990, pp. 575–631.
204. Rajeev Motwani and Prabhakar Raghavan. Randomized Algorithms. Cambridge University Press, Cambridge, 1995.
205. Rajeev Motwani and Prabhakar Raghavan. Randomized Algorithms. ACM Computing Surveys, vol. 28, no. 1, March 1996, pp. 33-37.
206. Leopoldo Nachbin. Topologia e Ordem. University of Chicago Press, Chicago, 1950, pp. 1-114.
207. Leopoldo Nachbin. Topology and Order. Van Nostrand, Princeton, 1965.
208. Michael A. Nielsen and Isaac L. Chuang. Quantum Computation and Quantum Information – the 10th Anniversary Edition. Cambridge University Press, Cambridge, 2010.
209. Tobias Nipkow, Lawrence C. Paulson, and Markus Wenzel. Isabelle/HOL – A Proof Assistant for Higher-Order Logic, Lecture Notes in Computer Science 2283, Springer, Berlin, 2002.
210. B. Nordström, K. Peterson, and J.M. Smith. Programming in Martin-Löf's Type Theory. The International Series of Monographs on Computer Science. Clarendon Press, Oxford, 1990.
211. James R. Norris. Markov Chains. Cambridge University Press, Cambridge, 1997.
212. Kristen Nygaard and Ole-Johan Dahl. The Development of the SIMULA Languages. The 1st ACM SIGPLAN Conference on History of Programming Languages, pp. 245–272. ACM Press, New York, 1978.
213. Chris Okasaki. Purely Functional Data Structures. Cambridge University Press, Cambridge, 1999, pp. 1–223.
214. Sungwoo Park. A Calculus for Probabilistic Languages. In Z. Shao, P. Lee, eds.: Proceedings of TLDI'03: The 1st ACM SIGPLAN International Workshop on Types in Languages Design and Implementation, ACM Press, New York, 2004, pp. 38–49.
215. Sungwoo Park. A Programming Language for Probabilistic Computation. Ph.D. Thesis, CMU-CS-05-137, Carnegie Mellon University, August 2005.

216. Terence Parr, Sam Harwell, and Kathleen Fisher. Adaptive LL($\star$) Parsing – the Power of Dynamic Analysis. In: Proceedings of OOPSLA 2014 – 28th ACM Conference on Object-Oriented Programming, Systems, Languages & Applications, ACM Press, New York, 2014, pp. 579–598.
217. Simon L. Peyton Jones. The Implementation of Functional Programming Languages, Prentice Hall, Upper Saddle River, May 1987, pp. 1–500.
218. Avi Pfeffer. IBAL – A Probabilistic Rational Programming Language. In: Proceedings of IJCAI'01 – the 17th International Joint Conference on Artificial Intelligence, vol. 1, Morgan Kaufmann, San Francisco, 2001, pp. 733–740.
219. Benjamin C. Pierce. Types and Programming Languages. MIT Press, Cambridge, 2002.
220. Gordon D. Plotkin. Call-by-Name, Call-by-Value and the λ-Calculus, Theoretical Computer Science, vol. 1, no. 3, 1975, pp. 125–159.
221. Gordon D. Plotkin. A Powerdomain Construction. In: SIAM (Society for Industrial and Applied Mathematics) Journal of Computing, vol. 5, no. 3, 1976, pp. 452–487.
222. Gordon D. Plotkin. A Powerdomain for Countable Non-Determinism. In M. Nielsen, E.M. Schmidt, eds.: Proceedings of ICALP'82: the 9th International Colloquium on Automata, Languages and Programming, LNCS 140, Springer, Berlin, 1982, pp. 412–428.
223. Gordon D. Plotkin. LCF Considered as a Programming Language. Theoretical Computer Science, vol. 5, no. 3, December 1977, pp. 223–255.
224. Gordon D. Plotkin. Domains – Pisa Notes on Domains. University of Edinburgh, 1983.
225. Michael O. Rabin. Probabilistic Automata. Information and Control, vol. 6, no. 3, 1963, pp. 230–245.
226. Michael O. Rabin. 1976. Probabilistic Algorithms. In J.F. Traub, ed.: Algorithms and Complexity – Recent Results and New Directions, Academic Press, Cambridge, 1976, pp. 21–39.
227. Michael O. Rabin. Probabilistic Algorithms for Testing Primality. Journal of Number Theory, vol. 12, no. 1, 1980, pp. 128–138.
228. Norman Ramsey and Avi Pfeffer. Stochastic Lambda Calculus and Monads of Probability Distributions. In: J. Launchbury, J.C. Mitchell, eds.: Proceeding of POPL 2002 – the 29th SIGPLAN-SIGACT Symposium on Principles of Programming Languages, ACM SIGPLAN Notices, vol. 37, no. 1, January 2002, pp. 154–165.
229. Manfred Reichert and Peter Dadam. ADEPTflex – Supporting Dynamic Changes of Workflows Without Losing Control. Journal of Intelligent Information Systems, vol. 10, no. 2, 1998, pp. 93–12.
230. Dan Romik. MATH/STAT 325a – Probability Theory Lecture Notes. University of California Davis, Fall 2011.
231. Stuart Russell. Unifying Logic and Probability. Communications of the ACM, vol. 58, no. 7, July 2015, pp. 88–97.
232. Nasser Saheb-Djahromi: Probabilistic LCF. In J. Winkowski, ed.: Proceedings of MFCS 1978 – the 7th Symposium on Mathematical Foundations of Computer Science, Lecture Notes in Computer Science 64, Springer, Berlin, 1978, pp. 442–451.
233. Nasser Saheb-Djahromi. CPO'S of Measures for Nondeterminism. Theoretical Computer Science, vol. 12, no. 1, 1980, pp. 19–37.

234. Eric Schechter. Handbook of Analysis and Its Foundations. Elsevier, Amsterdam, 2015.
235. David A. Schmidt. Denotational Semantics – A Methodology for Language Development. Allyn & Bacon, Boston, August 1986.
236. Moses Schönfinkel. Über die Bausteine der mathematischen Logik. Mathematische Annalen, no. 92, 1924, pp. 305–316.
237. Dana S. Scott. A Type-Theoretic Alternative to ISWIM, CUCH, OWHY. Theoretical Computer Science, vol. 121, no. 1–2, 1993, pp. 411–440.
238. Dana S. Scott. Outline of a Mathematical Theory of Computation. Technical Monograph PRG-2, Oxford University Computing Laboratory, Oxford, November 1970.
239. Dana S. Scott. Continuous Lattices. Technical Monograph PRG-7, Oxford University Computing Laboratory, Programming Research Group, August 1971.
240. Dana S. Scott. Data Types as Lattices. Society for Industrial and Applied Mathematics (SIAM) Journal on Computing, vol. 5, no. 3, pp. 522–587, 1976.
241. Dana S. Scott. Domains for Denotational Semantics. In M. Nielsen, E. M. Schmidt, eds.: Proceedings of ICALP'82 – the 9th Colloquium on Automata, Languages and Programming, LNCS 140, Springer, Berlin, pp. 577–613.
242. Dana S. Scott. Stochastic λ-Calculi – An Extended Abstract. Journal of Applied Logic, vol. 12, no. 3, 2014, pp. 369–376.
243. Dana S. Scott and Cristopher Strachey. Toward a Mathematical Semantics for Computer Languages. Technical Monograph PRG-6, Oxford University, 1971.
244. Koushik Se, Mahesh Viswanathan, and Gul Agha. Vesta: A Statistical Model-Checker and Analyzer for Probabilistic Systems. In: QEST 2005 – Proceedings of the 2nd International Conference on the Quantitative Evaluation of Systems, IEEE Computer Society, Los Alamitos, 2005, pp. 251–252.
245. Richard Serfozo. Basics of Applied Stochastic Processes – Probability and its Applications. Springer, Berlin, 2009.
246. Michael B. Smyth. Power Domains. Journal of Computer and System Sciences, vol. 16, no. 1, 1978, pp. 23–36.
247. Michael B. Smyth. Power Domains and Predicate Transformers – A Topological View. In J. Diaz, ed.: Proceedings of ICALP '83 – the 10th International Colloquium on Automata, Languages and Programming, LNCS 154, Springer, Berlin, 1983, pp. 662–676.
248. Robert M. Solovay and Volker Strassen. A Fast Monte-Carlo Test for Primality. SIAM Journal on Computing, vol. 6, no. 1, 1977, pp. 84–85.
249. Sam Staton, Hongseok Yang, Chris Heunen, Ohad Kammar, and Frank Wood. Semantics for Probabilistic Programming, Higher-Order Functions, Continuous Distributions and Soft Constraints. In: Proceedings of LICS 2016 – the 31st Annual ACM/IEEE Symposium on Logic in Computer Science, ACM, New York, 2016.
250. Sam Staton, Hongseok Yang, Chris Heunen, Ohad Kammar, and Frank Wood. Semantics for Probabilistic Programming, Higher-Order Functions, Continuous Distributions and Soft Constraints, version 3. In: The Computing Research Repository, arXiv:1601.04943, Cornell University Library, May 2016, pp. 1–10.

251. David Stirzaker. Elementary Probability. Cambridge University Press, Cambridge, 2010.
252. Anatoli V. Skorohod. Limit Theorems for Stochastic Processes. Theory of Probability & Its Applications, vol. 1, no. 3, 1956, pp. 261–290.
253. Joseph E. Stoy. Denotational Semantics. The Scott-Strachey Approach to Programming Language Theory. MIT Press, Cambridge, 1981.
254. Daniel W. Stroock. An Introduction to Markov Processes. Springer, Berlin, 2005.
255. Alfred Tarski. A Lattice-Theoretical Fixpoint Theorem and its Applications. Pacific Journal of Mathematics, vol. 5, no. 2, 1955, pp. 285–309.
256. T. Thalhammer, M. Schrefl, and M. Mohania. Active Data Warehouses – Complementing OLAP with Analysis Rules. Data & Knowledge Engineering 39, 2001, pp. 241–269.
257. The MathWorks Inc. Simulink Reference – Matlab & Simulink. The MathWorks Inc., 2015.
258. Regina Tix, Klaus Keimel, and Gordon D. Plotkin. Semantic Domains for Combining Probability and Non-Determinism. Electronic Notes in Theoretical Computer Science, vol. 222, Elsevier, 2009, pp. 3–99.
259. Steven Vickers. Information Systems for Continuous Posets. Theoretical Computer Science, vol. 114, no. 2, June 1993, pp. 201–229.
260. Barbara Weber and Manfred Reichert. Enabling Flexibility in Process-Aware Information Systems – Challenges, Methods, Technologies. Springer, Berlin, 2012.
261. Gerhard Weiss (editor). Multiagent Systems, 2nd edn. MIT Press, Cambridge, April 2013.
262. B.A Wichmann, A.A. Canning, D.L. Clutterbuck, L.A. Winsbarrow, N.J. Ward, and D.W.R. Marsh. Industrial Perspective on Static Analysis. Software Engineering Journal, Mar 1995, pp. 69–75.
263. Glynn Winskel. The Formal Semantics of Programming Languages – an Introduction. MIT Press, Cambridge, February 1993.
264. Andrew Chi-Chin Yao. Probabilistic Computations – Toward a Unified Measure of Complexity. In: Proceedings of SFCS'77 – the 18th Annual Symposium on Foundations of Computer Science, IEEE Computer Society, Los Alamitos, 1977, pp. 222–227.
265. Andrew Chi-Chin Yao and F. Frances Yao. On the Average-Case Complexity of Selecting the kth Best. Proceedings of FOCS'78, 19th Annual Symposium on Foundations of Computer Science, IEEE Computer Society 1978, Los Alamitos, pp. 280–289.
266. Andrew Chi-Chin Yao and F. Frances Yao. On the Average-Case Complexity of Selecting the kth Best. Society for Industrial and Applied Mathematics (SIAM) Journal on Computing, 1982, vol. 11, no. 3, pp. 428–447.
267. Hakan L. S. Younes. Ymer: A Statistical Model Checker. In Kousha Etessami, Sriram K. Rajamani, eds.: Proceedings of CAV'2005 – 17th International Conference on Computer Aided Verification, LNCS 3576, Springer, 2005, pp. 429–433
268. Moshe Zukermann. Introduction to Queueing Theory and Stochastic Teletraffic Models. In: The Mathematical Research Repository, arXiv:1307.2968, Cornell University Library, January 2016.
269. `https://github.com/tyler-barker/Randomized-PCF`

Index

Symbols

P

Q

R

S

Z

MIX
Papier aus verantwortungsvollen Quellen
Paper from responsible sources
FSC® C105338

If you have any concerns about our products,
you can contact us on
ProductSafety@springernature.com

In case Publisher is established outside the EU,
the EU authorized representative is:
Springer Nature Customer Service Center GmbH
Europaplatz 3, 69115 Heidelberg, Germany

Printed by Libri Plureos GmbH
in Hamburg, Germany